Mixed-Gender Basic Training

The U.S. Army Experience, 1973–2004

by

Anne W. Chapman

U.S. Army Training and Doctrine Command
Fort Monroe, Virginia, 2008

Library of Congress Cataloging-in-Publication Data

Chapman, Anne W.
  Mixed-gender basic training : the U.S. Army experience, 1973–2004 /
Anne W. Chapman.
      p. cm.
  Includes bibliographical references and index.
  1. Basic training (Military education)—United States. 2. United
States. Army Training and Doctrine Command)—History. 3. United States.
Army)—Women. 4. Women soldiers)—Training of)—United States. 5.
Coeducation)—United States)—History. I. Title.
  U408.3.C44 2008
  355.5'4097309045)—dc22

                                                        2008034776

For sale by the Superintendent of Documents, U.S. Government Printing Office
Internet: bookstore.gpo.gov   Phone: toll free (866) 512-1800;   DC area (202) 512-1800
Fax: (202) 512-2104 Mail: Stop IDCC, Washington, DC 20402-0001

ISBN 978-0-16-079420-9

Contents

Foreword

• •

The U.S. Army Training and Doctrine Command (TRADOC) originated in 1973, the same year the American military became an all-volunteer force. The primary mission of TRADOC then, as now, was to train and educate competent and adaptive soldiers and leaders at all levels. Later in that same decade, the Women's Army Corps ceased to exist, and women integrated with men in training and education. The simultaneous development of TRADOC, the all-volunteer Army, and gender integration irrevocably changed TRADOC and the Army. Research historian Anne Chapman, Ph.D., does an excellent job of charting TRADOC's responses to changing training requirements and approaches during the command's first three decades.

Intensified recruitment brought increasing numbers of women into the Army. Basic training for men and women came under one management structure, TRADOC, for the first time. As a result, the command wrestled with complex training issues: Should men and women be trained together? Given the legal prohibition against women going into combat, combined with the changing nature of combat, which skills should be gender-neutral and which should be segregated?

At the heart of these questions is the relationship between the American military and the society it defends. The increasing role of women in the military parallels the same trend in other professions throughout the nation. The Army is, however, a mission-driven organization, not a catalyst for social change. The people of the United States expect the Army to train and field land forces capable of defending the country's interests, and young men and women have answered the call to duty. It is critical to understand the institutional history of the Army's approach to training these young Americans to be soldiers, and the influence of gender on training, as we continue to wrestle with many of these same issues into the future.

The Army will continue to teach its values and the warrior ethos, instill discipline, and provide the skills needed by all soldiers to contribute immediately to their first unit of assignment. TRADOC can do no less.

Fort Monroe, Virginia
September 2008

ANTHONY R. JONES
Lieutenant General, U.S. Army
Acting Commanding General
U.S. Army Training and Doctrine
Command

Preface

● ●

Since the advent of the all-volunteer force in 1973, a major effort for the U.S. Army has been the configuration of a force increasingly dependent on female soldiers. Not the least of the problems demanding solution was the controversy stirred by the integration of men and women recruits in basic combat training. The issues were numerous, complex, and tenacious as the nation's largest military service sought to design a basic training program that met both the goal of military readiness and the increasing demands of women for equality of opportunity and treatment. This volume is an account of the many currents, some ongoing, that informed the Army's struggle to design a basic training course acceptable to the nation's civil and military leadership, the general public, various special interest groups, and the young men and women undergoing their first experience as soldiers.

This study employs a mixture of topical and chronological organization. Although there is brief attention to the long-range historical perspective, the major focus is on the period from 1973 to 2004. The author's aim is to tell the Army's story of mixed-gender training at the initial-entry level. There is no attempt to suggest solutions for past or present problems.

This volume is the result of a wide variety of sources. Perhaps most important are the multitude of studies conducted by the Army itself. There are also the reports of several congressional committees and hearings in both houses of the U.S. Congress. Articles from the print media are numerous, especially in the 1976–1981 and 1994–1998 periods. The reason for this concentration will become clear to the reader. There is also an extensive body of secondary works on the subject in which authors tend to draw heavily on each other's work and conclusions. Almost without exception, these writings take one of two approaches: (1) a focus on the dangers of women in the military to defense preparedness and readiness or (2) the argument that women cannot achieve full citizenship until they receive equal opportunity in the military workplace. This study is an effort to avoid this polarity and to tell what happened, why, and how it happened, not what could or should have happened.

Most of the primary sources cited are located in the TRADOC Historical Research Collection in the TRADOC Military History Office in Fort Monroe, Va. Some records of congressional hearings are available on the Library of Congress's Web site. Most articles are also available on the Internet.

The author owes a large debt of gratitude to a number of people who believed in the project to record the Army's efforts to make a major change in the way new soldiers are trained. Historians on the staff of the Military History Office offered encouragement, support, and understanding, especially when other major projects took precedence. Special thanks are owed to Steve Small, Ph.D., of the Army's Picatinny Arsenal, for sharing his personal and scholarly knowledge of the subject and for suggestions on portions of the manuscript. The TRADOC Technical Library staff was always willing to provide whatever support was needed, as well as numerous articles, Defense Technological Information Center documents, and volumes through the Interlibrary Loan system. Special mention is due to the late Dr. Brooks E. Kleber, former TRADOC chief historian, who developed sufficient interest in the subject of women in the military to save many documents that otherwise might have been lost. Invaluable were the studies of Lt. Col. Mattie E. Treadwell USA (Ret.) of the Office of the Chief of Military History and Col. Bettie J. Morden USA (Ret.) of the Center of Military History for providing a solid base for this study.[1]

Many friends, colleagues, and participants in the gender integration of basic training contributed to the improvement of this study with information and suggestions. These include Gen. Paul F. Gorman USA (Ret.), former TRADOC deputy chief of staff for Training; Lt. Gen. Robert H. Forman USA (Ret.), former TRADOC deputy commanding general for Training; Gen. William W. Hartzog USA (Ret.), former TRADOC commanding general; Brig. Gen. Evelyn P. Foote USA (Ret.),

---

[1] Treadwell, *The Women's Army Corps*, United States Army in World War II, Special Studies (Washington, D.C.: U.S. Army Center of Military History, 1954), publication available online at http://www.army.mil/cmh-pg/books/wwii/Wac/index.htm; Morden, *The Women's Army Corps, 1945–1978* (Washington, D.C.: U.S. Army Center of Military History, 1990, updated 2001), publication available online at http://www. army.mil/cmh-pg/books/wac/index.htm#contents.

a veteran of Pentagon panels, congressional hearings, and former commander of a mixed-gender training battalion; Judy Bellafaire, Ph.D., chief historian of the Women in Military Service Memorial Foundation; Capt. Lory Manning USN (Ret.), director of the Women's Research and Education Institute; and Judith Hicks Stiehm, author of several studies of gender-integrated training and former member of the Defense Advisory Committee on Women in the Service. Whatever flaws or shortcomings remain are the author's responsibility alone.

Fort Monroe, Virginia
September 2008

ANNE W. CHAPMAN, Ph.D.
Research Historian (Ret.)
Military History Office
U.S. Army Training and Doctrine
    Command

Introduction

• •

This is a study of the U.S. Army's experience in training enlisted men and women together for basic combat skills. It does not focus on the end of the draft or the all-volunteer Army. It is not a treatise on combat exclusion or women in combat. The focus is not on the women's liberation movement, the Equal Rights Amendment, or sexual harassment. Nor is it a story of the other U.S. military services' efforts to develop programs of mixed-gender training. However, the issues of training women for basic combat did not play out in a vacuum. Each of these interrelated currents, debates, and controversies helped to define and influence the issues and helped to shape the Army's new program. Collectively, they provided the background against which the Army had to make decisions. The corresponding activities by the Navy, Marine Corps, and Air Force are addressed only when they affected the Army decision-making process. Further, this study does not address the Army Nurse Corps, except in passing, because the experience of medical personnel does not parallel that of other female trainees.

As the Army sought to satisfy many interests in designing a mixed-gender training program, a pattern developed. From 1973, when military conscription officially ended, until 1978, the Army conducted numerous studies in an effort to identify actions that would be necessary for the establishment of a mixed-gender training program for new recruits. Finally, in 1978, with the elimination of the Women's Army Corps, a company-level program was put in place. That program encountered difficulties, including the failure of the states to ratify the proposed Equal Rights Amendment, which was quietly abandoned in early 1982.

For the next decade, public forums continued to demand either the unequivocal end of attempts at the cotraining of men and women or the immediate reinstitution of gender-mixed programs, which were perceived as offering women more opportunity. That stalemate ended with the insistence of Secretary of Defense Les Aspin that women be allowed greater opportunity, based on their relatively heavy and much-heralded participation in Operation DESERT SHIELD and Operation DESERT STORM, and on resource constraints, and interservice rivalries. In 1994, the

Army announced a basic training program that integrated women much more completely than previous attempts.

During the next two years, the new program appeared to be a qualified success as problems were addressed and, in general, solved. Only the issue of women in combat remained consistently in the limelight. Then, in 1996, a number of incidents of alleged sexual harassment and sexual abuse occurred at the Army's Aberdeen Proving Ground in Maryland and at other sites.[2] Those events convinced even more observers that mixed-gender training could be detrimental to women trainees and to military readiness. In the next three years, two major commissions and numerous congressional hearings studied the issue of gender-integrated training with an eye to the wisdom of continuing the practice or of eliminating it in all the services. The most recent commission reluctantly endorsed such training, with caveats, but that report did nothing to deter the strong movement among skeptics to study the effects of cotraining of men and women yet again.[3]

Remarkably similar issues arose each time the Army worked to design a new mixed-gender military training program for new recruits. The question of physical strength and endurance was always the most controversial, followed closely by the question of the assignment of women and whether the public would accept women in combat. Pregnancy, damage to the male ego and to male bonding, emotions and aggressiveness, and fraternization also played roles in the debate.

In this study, a number of terms describe the training of men and women together in U.S. Army basic training. From 1994 to 2003, *gender-integrated training* was the term most often used by the Army, the public, and the press. A variety of terms, such as *coed* and *unisex*, have also been used. On occasion, the word *integration* appears alone, but this practice should not be regarded as a reference to racial arrangements.

---

[2] These incidents involved trainees and training cadre in advanced individual training, not basic combat training.

[3] The 1999 Blair Commission is discussed in Chapter IV of this work. More recent events are covered in Chapter V.

Mixed-Gender Basic Training

The U.S. Army Experience, 1973–2004

I

Background

• •

This initial chapter on the evolution of gender-integrated basic training provides a brief overview of the history of women in the U.S. Army.[1] It also presents, in capsule form, some of the major social debates and controversies of the mid- to late twentieth century concerning women in the military services. These debates went far in determining the approaches taken by the Army in establishing basic training units that included both genders. It is never wise for an institution to ignore its own past, and the Army's experience with women in the military is no exception.

A wide variety of controversial topics and currents, individually and collectively, constituted the background against which the U.S. Army addressed the issues of women in the Army, in general, and the training of men and women together during their initial introduction to the service, in particular. In addition, the existing social and political climate played a role in defining the issues that the Army had to consider in its effort to design a mixed-gender curriculum for its basic combat training (BCT) program. During the period from 1973 to 2004, those issues remained remarkably consistent.

A Historical Overview: The American Revolution to 1973

Before World War II, service in the U.S. military was primarily a masculine calling, as it was in the armed forces of most nations. In general, only necessity overrode that tradition, although women have participated in military actions and undertaken military roles in American history on a number of occasions since colonial times. Between 1776 and 1918, women served as cooks, nurses, and medical assistants;

---

[1] For a recent report on this and other gender-related issues that includes an annotated bibliography, see Army Research Institute for the Behavioral and Social Sciences, "Women in the U.S. Army: An Annotated Bibliography," Apr 2002.

laundry and armament workers; and even, sometimes, as combatants. Some stories have even been told about a few women, generally disguised as men, who took up arms.

In General George Washington's army, women cooked, chopped firewood, built shelters, and nursed sick and wounded soldiers.[2] Women's roles during the Civil War are well documented. Many served as nurses under the direction of Clara Barton, the founder of the American Red Cross. Mary Edwards Walker received a commission in the Union Army as a medical doctor and was awarded the Congressional Medal of Honor. Thousands of other women supported the military-industrial base by working in government-owned arsenals and armories.[3]

By the turn of the twentieth century, women were gaining momentum in their efforts to serve. During the Spanish-American War in 1898, when an epidemic of typhoid fever created a severe shortage of male nurses, more than 1,500 women served as military nurses under contract. Army and Navy leadership took steps to create a permanent nurse corps largely as a result of the service of these women. The U.S. Congress established the Army Nurse Corps in 1901 and the Navy Nurse Corps in 1908 as auxiliaries of these military services.[4]

World War I proved to be a watershed conflict for military women—except those in the Army. In 1917, the Naval Reserve began to recruit women as yeomen. During the war some 12,500 "yeomanettes," as they were popularly called, served on active duty as clerks, secretaries, and nurses and, increasingly, as draftsmen, translators, and recruiters. Shortly before the end of the war, the Marine Corps recruited several hundred female Marine reservists to make up for manpower shortfalls. Nonetheless, the War Department (which later became the Department of the Army) continued to prohibit the enlistment of women into the military.[5] The Army did, however, employ on contract more than 200 French-speaking American women in France as switchboard operators. When the war ended, the women were not granted honorable discharges or veterans' benefits because they had not served in uniform.

---

[2] William W. Fowler, *Frontier Women: An Authentic History of the Courage and Trials of the Pioneer Heroines of Our American Frontier* (Stamford, Conn.: Longmeadow Press, 1995), pp. 136–37.

[3] Stephen Small, "Women in American Military History, 1776–1918," *Military Review* LXXVIII (Mar-Apr 1998): 101–04.

[4] Ibid.

[5] Maj Gen Jeanne Holm USAF (Ret.), *Women in the Military: An Unfinished Revolution*, rev. ed. (Novato, Calif.: Presidio Press, 1992), pp. 10, 12–13. Even Gen Pershing protested this prohibition, to no avail.

Not until 1977 were these women granted veteran status under the GI Bill Improvement Act of that year.[6]

The two decades following World War I were generally a stagnant period for women who sought to perform military service. In 1920, the director of Women's Programs for the Army, Anita Phipps, proposed the creation of a permanent Women's Service Corps whose members would receive full military salaries. With the Army shrinking and national interest in military affairs waning, however, the War Department rejected the director's idea, and her position was eliminated in 1931.[7] Several other proposals to raise a force of women to be used in time of war were similarly rejected. Given the moribund state of the U.S. Army between the two World Wars, little attention was paid to the need to plan for the use of women in the Army in any future conflicts.[8]

World War II brought severe manpower shortages that forced the services to consider filling some support positions with women, thereby "freeing a man to fight." In May 1942, the Army formed the Women's Army Auxiliary Corps (WAAC), which allowed women to serve *with* but not *in* the Army.[9] The WAAC's status as an auxiliary meant that women did not receive pay equal to that of men, did not have the same ranks, and were not entitled to full benefits. Those problems were addressed and largely remedied by creation of the Women's Army Corps (WAC) in July 1943 and the dissolution of the WAAC at the end of September in that same year. The WAC director reported directly to Army Chief of Staff General George C. Marshall. By the end of the war, approximately 400,000 women had served in the military.[10]

Immediately after World War II, the U.S. military rapidly demobilized, falling from a force of about 12 million in 1945 to about 1.4 million by 1948. Correspondingly, the number of women in the military declined from 266,000 to about 14,000.[11] When it appeared that the

---

[6] *U.S. Statutes at Large* 95-202, sec. 401 (1977).

[7] Holm, *Women in the Military*, pp. 17–18.

[8] Forrest C. Pogue, *George C. Marshall: Organizer of Victory, 1943–1945* (New York: Viking Press, 1973), pp. 104–05.

[9] For a discussion of the creation of the WAAC and the WAC, see Lt Col Mattie E. Treadwell USA (Ret.), *The Women's Army Corps,* United States Army in World War II, Special Studies (Washington, D.C.: U.S. Army Center of Military History, 1954), pp. 24–45, 256–68.

[10] "Statistics on Women in the Military" (Washington, D.C.: Women in Military Service for America Memorial Foundation, Inc., revised 17 Jul 2006).

[11] Department of Defense, *Selected Manpower Statistics* (Office of the Assistant Secretary of Defense [Comptroller], Directorate for Information, Operation, and Control, May 1975 [processed]), pp. 22, 46; Martin Binkin and Shirley J. Bach, *Women and the Military* (Washington, D.C.: The Brookings Institution, 1977), p. 10.

authorization to continue the WAC would end in 1948, about half of the women remaining resigned. But a number of forces converged to ensure that women maintained some place in the peacetime military. Not the least of these forces was the concern that without conscription, which was due to lapse in March 1947, the armed services would not be able to meet their recruitment needs. Allowing women in the regular military, the Pentagon argued, would provide an additional source of personnel and a trained nucleus from which to expand the military in a national emergency. In addition, the institutionalizing of women's roles would provide a laboratory for determining how best to use women in the military.[12]

In June 1948, women were assured of a right to serve by the passage of the Women's Armed Services Integration Act (PL 80-625), signed by President Harry S. Truman. The new law was designed to provide for mobilization of women in the event of general war. Debate in Congress was not unduly heated and generally focused on two principal issues: First, should women become a part of the regular military establishment or be maintained in a reserve status? Second, how many women should be allowed to serve at a given time?[13] The first question was quickly decided in favor of regular status. On the second, the Pentagon prevailed. A 2 percent ceiling was placed on the proportion of enlisted women in the services; female officers would not exceed 10 percent of female enlisted strength, not including nurses. Of the services, only the Army retained women in a separate corps, the WAC. All Army women who were not members of the Nurse Corps or Women's Medical Specialist Corps were to be members of the WAC, which would be responsible for most career issues, including training and promotions.[14]

The 1948 legislation represented a major advance for women in the services, but it also institutionalized some limitations on enlistment, promotion, and benefits. Further, it restricted the assignment of women to positions that would not expose them to direct combat. Women in the Air Force could not be assigned to aircraft on combat missions. In the Navy, women could be assigned only to transports and hospital ships. The Army was given the authority to establish policy for the employ-

---

[12] Binkin and Bach, *Women and the Military*, p. 10; House Armed Services Committee, Subcommittee on Organization and Mobilization, *Hearing on S. 1641 to Establish the Women's Army Corps in the Regular Army . . .*, 80th Cong., 1st sess., 1948, 5595.

[13] Binkin and Bach, *Women and the Military*, p. 11.

[14] Vicki L. Friedl, comp., *Women in the United States Military, 1901–1995: A Research Guide and Annotated Bibliography* (Westport, Conn.: Greenwood Press, 1996), p. 99.

ment of women according to its needs. In the absence of law, that service adopted its own policy of combat exclusion.[15]

The role of women in the military in the two decades after World War II was severely limited. In 1950, only 22,000 women were on active duty. Manpower pressure during the Korean War generated attempts to enlist larger numbers of women, but these efforts largely failed, with only some 49,000 women serving at the peak of the war. Even those servicewomen were generally restricted to traditionally female occupations in the clerical and health care arenas.[16]

By the beginning of the Vietnam War, the total number of women on active duty had shrunk to 30,600. In 1966, however, facing an increasing demand for manpower, the Department of Defense established an Inter-Service Working Group on Utilization of Women in the Armed Services to assess the role of women in the military.[17] Partly as a result of that study, the 2 percent ceiling on female enlisted strength, enacted in 1947, was eliminated; the barrier to promotion to general officer was struck down; and the existing differences between retirement provisions for men and women were removed (PL 90-130). The combat exclusion clause for women in the Navy and Air Force remained. Despite these gradual improvements, women continued to make up less than 2 percent of total military strength throughout the 1960s. Approximately 6,000 to 12,000 women served in Vietnam.[18]

Military, Social, and Political Currents

The limited experience of women in the U.S. military during the first two centuries of the nation's existence remained in the background of the debates that took place throughout the mid- to late twentieth century. No discussion of the integration of women with men into the same basic training units would stray far from the fundamental questions of the role of women in the Army: What would that role be? For what positions or specialties would women receive training? What was the role of women, if any, in combat units? How many women should be in the military in the first place? What differences between men and women

[15] Ibid. The reinstitution of conscription two weeks after the enactment of the 1948 Women's Armed Services Integration Act removed the services' concerns about manpower shortages. See also Binkin and Bach, *Women and the Military*, p. 11.

[16] Binkin and Bach, *Women and the Military*, pp. 11–12; Friedl, *Women in the United States Military, 1901–1995*, p. 12.

[17] Binkin and Bach, *Women and the Military*, p. 12.

[18] Binkin and Bach, *Women and the Military*, p. 12; Friedl, *Women in the United States Military, 1901–1995*, p. 12. Of that total, only approximately 700 were WACs.

had to be accommodated, and what differences could be dismissed as merely cultural or superficial? These questions had an impact on the U.S. Army's decisions over time concerning the integration of men and women in basic training. These entangled and interrelated debates took place in a multitude of forums from 1973 to 2004 and had a major impact on the changing role of women in the Army. The key event that precipitated much of the debate was the termination of conscription and the creation of the all-volunteer Army in 1973.

The All-Volunteer Army and the Numbers

In 1972, President Richard M. Nixon ran for reelection on a platform that included ending conscription for the military forces. The Defense Department ended draft calls in January 1973, six months before Congress allowed induction authority to expire on 1 July 1973. With that action, the United States became the first nation to attempt to field a military force of more than two million that depended only on volunteers. Both Congress and the Department of Defense realized that the expanded recruitment and enlistment of women would be essential to maintaining the size of the military, especially the Army, which relied heavily on conscripts until 1973.

In early 1973, women made up about 2.4 percent of the Army's total strength, but that number grew to 6.7 percent by 1977. Three years later, the number of women in the Army reached approximately 10 percent.[19] During the administration of President James E. Carter (1977–1981), who strongly supported women in the military, the number of women on active duty across the services rose to 173,450 in January 1981 from approximately 45,000 in January 1972.[20]

None of this growth in the number of women in the Army was achieved without great uncertainty and many misgivings. Brig. Gen. Edith Foote USA (Ret.), who commanded a gender-mixed battalion, noted: "The training base and supply system were totally unprepared for the greatly increased numbers of women to train. The Army 'cobbled' together multiple battalions and mixed cadres to train women. [The service] was woefully incapable of clothing, equipping, and housing women in the Army for years. Further, the Army failed almost totally in preparing its men and women for service in a much more 'integrated' mode."[21]

---

[19] Russell F. Weigley, *History of the United States Army,* enl. ed. (Bloomington: Indiana University Press, 1984), p. 567.

[20] Holm, *Women in the Military*, p. 387.

[21] Note from Brig Gen Edith Foote USA (Ret.) to the author, Dec 2004.

Senior defense leaders and members of Congress, publicly and privately, expressed grave concern that the rapid increase in the proportion of women in the military services would produce a corresponding decrease in military readiness and mission capability. The all-volunteer force (AVF) and the resultant increase in the number of women serving in the military were blamed, at least in part, for numerous shortcomings, including an increase in illegal drug use by service members, poor and broken equipment, low morale, poor training, and poor unit cohesion, to name a few. Many critics of the AVF saw reinstitution of conscription as a means of correcting many of these problems while reducing dependence on the recruitment of ever-greater numbers of females. When recruiting efforts fell short, criticism of the AVF became commonplace in the national press, yet the draft remained deeply unpopular with the American people.

These perceived problems, coupled with the arrival of the presidential administration of Ronald Reagan in 1981, led to a push by the Army to reexamine the issue of women soldiers. In testimony before the Senate Armed Services Manpower Subcommittee on 26 February 1981, Army officials announced that the service intended to hold female strength at 65,000 enlisted personnel until the role of women and their impact on readiness had been studied more comprehensively.[22] This initiative was short-lived, however. In August 1982, the Pentagon announced an increase in recruiting goals for women to 75,000 by 1989.[23]

The Women's Liberation Movement, the ERA, and DACOWITS

Students of social and political activism on behalf of women's rights usually speak of the movement that began in the early 1960s as the "second wave" of liberal feminism, reserving "first wave" status for the women's suffrage movement of the mid-nineteenth to the early twentieth century. From its beginnings in the turbulent era of the 1960s, this second wave of the women's movement grew slowly but steadily into the 1970s. Subsequently, a reaction against feminism began to build. When conservatives claimed the presidency with the election of Ronald Reagan in 1980, feminists were put on the defensive. Many believed that the women's movement stalled during the Reagan years as it struggled to hang on to its perceived gains. The 1990s and the opening years of the twenty-first century saw a number of changes in

---

[22] Holm, *Women in the Military*, pp. 388–89.
[23] Weigley, *History of the United States Army,* p. 568.

the feminist movement, along with a shifting focus to issues of women in the workplace.[24]

Most American feminist groups have not paid as much attention to issues of women in the military as might be expected. Some Washington-based women's advocacy organizations with national agendas, such as the National Organization of Women, the Women's Research and Education Institute, and the Women's Equity Action League, have occasionally addressed some issues related to military service, including sexual harassment and "masculinized divisions of labor." In large part, however, these organizations are not focused on issues related to a gender-integrated military.[25] The reasons for this inattention are complicated but identifiable.

First, the strong emergence of peace activism within the women's movement of the 1970s–1980s made the issue of women in the military seem out of place. The powerful pacifist branch of feminism certainly did not condone and sometimes outright condemned any participation of women in the military. Many feminists abhorred the combat role of the military profession and disagreed deeply with the foreign policy that members of the military were sworn to serve. Second, women who enlisted in the military services tended to come from more traditional and conservative backgrounds and were generally not attracted to liberal feminist groups. As a result, many goals, beliefs, and agendas of the feminist movement contradicted or differed dramatically from those of military women. While the women's movement campaigned to win equal pay for equal work, military women already possessed a system that paid by rank, not by job. Further, many feminists believed that the only position consistent with gender equity was to support the inclusion of women in the draft and in combat roles, a position many military women (and probably most American women) did not support. Despite the fact that the rise of feminism and women's integration into the military were, for the most part, separate events, military women have generally benefited from feminists' efforts, whether or not particular feminist groups were concerned with the military or consciously trying to raise the status of women in the armed forces. Affirmative-action and class-action lawsuits and their resultant political pressures helped to expand women's roles in the military and lift restrictions on their num-

---

[24] Flora Davis, *Moving the Mountain: The Women's Movement in America Since 1960* (Urbana and Chicago: University of Illinois Press, 1991, 1999), pp. 15–16.

[25] Cynthia H. Enloe, "The Politics of Constructing the American Woman Soldier," in *Women Soldiers: Images and Realities*, ed. Elisabetta Addis, Valeria E. Russo, and Lorenza Sebesta (New York: St. Martin's Press, 1994), pp. 81–110, quotation p. 90.

bers and promotion opportunities. And, increasingly, the courts applied rulings involving civilians to personnel policies in military settings.[26]

Next to the advent of the all-volunteer Army, no other single factor contributed more to the unexpected and unprecedented expansion of women's participation in the armed forces, as well as to questions about the impact of that participation, than the proposed Equal Rights Amendment (ERA). The National Woman's Party had introduced the ERA in Congress in 1923, only three years after passage of the Nineteenth Amendment granting women the right to vote. After nearly fifty years of debate, changes in wording, massive lobbying by women's organizations, and shifting pro-amendment alliances, Congress passed the amendment on 22 March 1972. If it had been ratified by three-fourths of the states, it would have become the Twenty-Seventh Amendment to the U.S. Constitution. During its movement through the Senate and House of Representatives, the issues of conscription and combat exclusion for women proved the most difficult for amendment supporters. Perhaps the most serious threat to passage of the proposed amendment in Congress was the introduction of two amendments by Senator Samuel Ervin Jr. of North Carolina that would have specifically excluded women from the draft and from combat.

The Ervin amendments were overwhelmingly defeated, but their defeat also signaled to some women that equal opportunity might come with equal responsibility as citizens, including conscription and involuntary assignment to combat units. Opponents of the ERA helped to defeat the Ervin amendments in the hopes that the possibility of drafting and deploying women in combat would, in turn, help to defeat the ERA. As for Congress, the rejection of the amendments suggested that a majority of its members anticipated an expanded role in the future for women in the military. As one commentator on the event noted, "The message beamed to the armed forces was that they would not be exempt from the mandate of the ERA if and when it became a part of the Constitution."[27] Note that in the late 1970s and into the 1980s, the services were firmly convinced that the ERA would be ratified by the states.[28]

As mentioned earlier, the feminist movement generally did not seem focused on efforts to improve the numbers and opportunities of women in the military. One organization, however, gave priority to the issue of women in the military and helped to compensate for the lack of

---

[26] Ibid., p. 89.

[27] Holm, *Women in the Military*, p. 264.

[28] Holm, *Women in the Military*, pp. 249, 264. For a full account of the history of the ERA, see also Davis, *Moving the Mountain*, pp. 29–45, 121–36, 385–411.

support from the women's rights movement. This organization was the Defense Advisory Committee on Women in the Services (DACOWITS), established in 1951 by then–Secretary of Defense General George C. Marshall. The organization had its genesis in the recommendation of a conference of civilian leaders held in June 1950 in the Pentagon to discuss the future utilization of women in defense. Several of the conference participants had served with Marshall and knew of his strong support for women in the military and his role in the establishment of the WAC during World War II; they recommended that a similar committee be formed to provide advice to the secretary. Further support for such a committee came in June 1951, after the beginning of the Korean War, from military leadership concerned with low recruitment rates for women.[29]

Originally, the mission of DACOWITS was to inform the public, particularly young women, of the value of military service and the opportunities available to them. In addition, the committee had the task of reassuring concerned parents that their daughters would be well supervised in the military. Overall, the goal was to improve the image and prestige of military women in the public mind as an aid to recruitment. Over time, that mission shifted to focus more strongly on how women were to be used in the military and on quality-of-life and legal issues perceived as having an impact on the readiness of military women.[30]

The committee, appointed by and reporting directly to the secretary of Defense, usually included thirty to forty civilian men and women—more often women—from academic, business, legal, artistic, and political professions. Each member served a three-year term on a rotating basis. The services appointed personnel to aid and advise the committee as military representatives, but they were not to be considered committee members. All DACOWITS members were required to visit military installations and facilities to talk with servicemen and servicewomen about their concerns and to solicit suggestions. The committee held semiannual conferences to receive responses to its requests for information from the services and Department of Defense components, discuss current issues and concerns, and formulate new requests before making policy recommendations to the secretary of Defense.

As an organization, DACOWITS had few connections with civil rights, peace, or feminist groups but on occasion engaged in its own brand of activism. Although the committee had only a limited impact

---

[29] Holm, *Women in the Military*, pp. 150–51.
[30] Ibid., p. 151.

in generating public interest in military women (its original charter), it enjoyed a political status that enabled it to command a hearing among Defense Department senior officials, congressional committees concerned with defense, and on occasion, the national press. A survey of the organization's recommendations since 1951 offers a timeline that reveals the peaks and valleys in the history of women in the military.

For almost twenty years following the Korean War, DACOWITS generally attracted little attention as the committee pursued quality-of-life issues for women in the military, such as assisting in efforts to obtain properly fitting uniforms and boots. The organization's major success during this time was the passage of PL 90-130 in 1967, which removed the ceiling on promotions for women in the military. After that, most members apparently shied away from controversial issues as the Vietnam War became increasingly unpopular.

The activities of DACOWITS increased during the mid-1970s and 1980s. Interest intensified in the 1990s as the Gulf War, the Navy Tailhook sexual abuse scandal, and sexual harassment charges at several Army installations attracted widespread public attention. Although DACOWITS was often characterized as a "radical feminist" organization, the record does not entirely support that label. The membership on numerous occasions was pulled in many directions.[31] On balance, however, the committee consensus usually came down on the side of change rather than maintenance of the status quo.[32]

Combat Exclusion and the Draft

The Women's Armed Services Integration Act of 1948 (PL 80-625) made women an integral part of the regular military establishment. This act also included a combat exclusion provision that stated: "Women may be assigned to all units except those with a high probability of engaging in ground combat, direct exposure to enemy fire, or direct physical contact with the enemy."[33] The law also specifically prevented women from serving on combat ships and aircraft.[34] For twenty-five

---

[31] One of the most well known members of DACOWITS was Sandra Day O'Connor, who served from 1974 to 1976 while a member of the Arizona state legislature. President Ronald Reagan nominated O'Connor, considered to be conservative, to the U.S. Supreme Court in 1981.

[32] DACOWITS recommendations since 1973 are available from the Defense Technological Information Center (DTIC) and its Internet site.

[33] Public Law 80-625, *Women's Armed Services Integration Act of 1948*, 62 Stat. 356–75.

[34] Ibid.

years, that legislation formed the background for often-heated debate within the military establishment concerning the appropriate assignments and utilization of women in the military.[35]

The combat exclusion provision applied only to the Air Force and Navy because disallowing women on certain ships or aircraft and, thus, keeping them out of combat was a relatively simple matter. Drawing the line for either the Army or the Marine Corps did not prove as easy. Neither service could produce an acceptable definition of *combat*. In the case of the Air Force and Navy, if Congress withdrew the combat exclusion provision, each service had the right to decide whether it wanted to open combat positions to women. The Army and Marine Corps, not specifically controlled by the law but adopting the spirit of it, kept women out of combat specialties as a matter of policy. In the case of the Army, the secretary had the authority to exclude women from combat pursuant to the law. As the issue of women in combat was repeatedly debated and pressure from feminist groups for change mounted, the service chiefs tended to look to congressional repeal of the law as a solution. At the same time, Congress was prohibited by the law and the "will of the American people," as it determined that will, from allowing women in combat.

For the Army, one of the greatest difficulties in defining where women could serve was the seemingly ever-changing definition of what constituted *combat*. *Combat* has traditionally been defined in terms of a unit's physical proximity to the enemy on the battlefield. As many more women became a part of the force and were allowed to serve in a greater number of military occupational specialties (MOSs) and as technology changed the nature of modern warfare, the battlefield itself became much more fluid. It soon became apparent that women serving in, for example, supply, repair, or communications units would be required to move into and out of battle zones to which they could not be assigned. In 1977, Secretary of the Army Clifford L. Alexander issued an updated combat exclusion policy: "Women may not serve in Infantry, Armor, Cannon Field Artillery, Combat Engineer, or Low Altitude Air Defense Artillery units of Battalion/Squadron or smaller size."[36]

---

[35] Those are the terms most often encountered regarding this issue in military accounts because the military occupational specialty (MOS) to which a female soldier was assigned often determined whether she was excluded from direct combat. The other services used different terminology. Media accounts of the debate almost universally employed the term *combat exclusion*.

[36] Cdr TRADOC [Starry] to distr, 040035Z Feb 1978, sub: Policy on Exclusion of Women from Combat; U.S. Code, Title 10, Sec. 3012.

In 1978, to further address the contradiction between policy and reality, a committee appointed by Army Chief of Staff General Bernard W. Rogers unofficially adopted a revised definition of *combat* in an attempt to clarify the issue: "Women will be excluded from positions which have as their primary function the crewing or operation of direct and indirect fire weapons."[37]

This revised definition replaced the position-on-the-battlefield concept with one that defined *combat* according to an individual's or unit's primary duty or mission. The 1978 working definition was replaced four years later during an extensive examination of all Army MOSs for the probability that the soldier would be involved in direct combat. The new definition—adopted from a Department of Defense definition of *close combat*—was more explicit: "Direct ground combat is engaging an enemy on the ground with individual or crew served weapons, while being exposed to direct enemy fire and to a high probability of direct physical contact with the enemy's personnel, and substantial risk of capture. Direct combat takes place while closing with the enemy by fire, maneuver, or shock effect in order to destroy or capture or while repelling assault by fire, close combat or counterattack."[38]

Meanwhile, the Army instituted a classification system to evaluate every MOS based on a complicated set of criteria, including unit mission, tactical doctrine, and location on the battlefield. (This Direct Combat Probability Coding [DCPC] System is discussed in more detail in Chapters II and III.) In 1988, the Department of Defense further confused the issue by developing a *risk rule* designed to standardize positions closed to women across the services: "Risks of exposure to direct combat, hostile fire, or capture are proper criteria for closing noncombat positions or units to women, providing that the type, degree, and duration of such risks are equal to or greater than that experienced by combat units in the same theater of operations."[39]

Over time, the Army, not bound by legal provisions in this matter, made minor changes to the definition of *combat*, and in general, those revisions made the definition more elastic to fit a rapidly changing situation.

In the same manner as the definition of *combat*, the assignment of women to occupational specialties received ever-increasing attention

---

[37] U.S. Army Administration Center, *Evaluation of Women in the Army*, Final Report, March 1978. This committee was appointed in August 1977.

[38] Department of the Army, Office of the Deputy Chief of Staff for Personnel, *Women in the Army Policy Review*, 12 Nov 1982, p. 7.

[39] As quoted in Carolyn Becraft, "Women in the Military, 1989–1990," Women's Research and Education Institute, Jun 1990.

as the number of women in the military services continued to rise. The Army alone sponsored at least ten major studies between 1972 and 1993 to determine how women could or should best be used.[40] Significant changes were made beginning in the early 1970s. Before a 1972 expansion of the number of women in the service, only 35 percent of all military enlisted job specialties were open to women. Following a reassessment during that year, more than 90 percent of military jobs were open to new recruits, regardless of gender.[41] Assignment to nontraditional jobs (not nursing or clerical duties) also increased rapidly. At least for the Army, however, these changes did not represent a smooth progression into the future.

For example, following a major study conducted in 1981 and 1982, Chief of Staff General Edward C. Meyer added twenty-three MOSs to the thirty-eight already classified as associated with combat.[42] Most of the twenty-three were restored beginning in 1983. Percentages, too, could be deceptive. For example, even though 50 percent of nontraditional MOSs were open to women, it did not follow that 50 percent of women were in nontraditional jobs because the MOSs that were closed included large numbers of specialties restricted to male personnel, especially in the Infantry and Armor branches.

The unresolved issue became even cloudier for the Army when, for the first time, gender-integrated units deployed to Panama during Operation JUST CAUSE in late 1989. The operation revealed confusion in the field concerning the definition of *combat* and highlighted some basic contradictions in Army policy related to the employment of women.

Shortly after the Panama deployment and with the support of DACOWITS, legislation was proposed to establish a trial program to test the suitability of women for the combat arms. In April 1990, the Army announced that it would not initiate such a program. In 1993, however, in the wake of the Persian Gulf War, the combat exclusion and assignment questions with regard to Air Force and Navy women, except for duty on submarines, were largely resolved when Secretary of Defense Les Aspin directed that women be allowed to compete for assignments in aircraft engaged in combat missions and aboard combat ships. On that occasion, the Army and Marine Corps were directed to study the possibilities of opening more assignments to women.

---

[40] Judith Hicks Stiehm, *Arms and the Enlisted Woman* (Philadelphia: Temple University Press, 1989), pp. 137–38.

[41] Ibid., p. 138.

[42] At the time, Meyer was attempting to comply with a new assignment system known as Direct Combat Probability Coding (DCPC).

None of the proposed legislation directly addressed the Army because that service maintained its combat exclusion policy by regulation, but the repeal of the laws governing the Air Force and Navy tended, some believed, to undermine Army policy. A controversial study in 1993, directed by the administration of President George H. W. Bush, failed to resolve the combat exclusion question for the Army, although some MOSs in Field Artillery and Air Defense Artillery were later opened to women. As of 2004, women continued to be denied assignment to Infantry, Armor, Combat Engineer, or Special Forces units. The issue continued to be contentious, and no matter the circumstances or the forum, the core questions of military readiness, cultural perceptions, and the rights of women as full citizens remained.

Whenever the issue of combat exclusion arose, concerns about registering women for the draft and actually drafting them followed close behind. If women were to be on the front lines, would or should they also be subject to conscription?[43] It is almost certain that most Americans never seriously considered such a move, not even in times of severe manpower shortages. The issue was kept constantly in the background, however, by the tentative insistence of some feminist groups that women would not achieve full citizenship until they were subject to the draft.

On 31 March 1947, as Congress debated the merits of adopting a system of universal military training, the Selective Service Act and its extensions, which had taken the United States through World War II, were allowed to lapse. To a government and public that envisioned war almost exclusively in terms of air and atomic power, the Army of the late 1940s seemed almost irrelevant to the communist challenge. In 1948, however, as tensions grew in Europe with the Russian blockade of Berlin, the communist coup in Czechoslovakia, and the Greek army's contest against Russian-sponsored communist guerrillas, interest in traditional forms of military power was renewed. As a result, Congress passed the Selective Service Act (PL 80-759) on 24 June 1948. This version of the draft would last for thirty-five years, through the Korean and Vietnam conflicts.

Just twelve days before the reinstatement of conscription, President Truman signed the Women's Armed Services Integration Act (PL 80-625). Despite efforts to the contrary, the law contained no prohibition against drafting women. With the males-only draft

---

[43] Sara Ruddick, "Drafting Women: Pieces of a Puzzle," in *Conscripts and Volunteers: Military Requirements, Social Justice, and the All-Volunteer Force,* Maryland Studies in Public Philosophy, ed. Robert K. Fullinwider (Lanham, Md.: Rowman and Littlefield, 1983), p. 214.

reinstated and low recruiting goals for women, the issue faded away, awaiting another manpower crisis. That crisis came in the spring of 1950 when North Korean forces invaded South Korea. Despite warnings from the National Security Resources Board Subcommittee on Manpower that a draft might be required for all young men and women if a military force of three million was required, that necessity never arose.[44] After Korea, any thoughts of drafting women essentially disappeared as the nation adopted a defense strategy based on massive retaliation and general war in which nuclear exchange would rapidly decide the outcome. Senior leaders in all the services even suggested that women should be eliminated from the armed forces as completely unnecessary to national defense.[45] Nor did the "flexible response" doctrine of the presidential administrations of John F. Kennedy and Lyndon B. Johnson produce any call for registering or drafting women.

As mentioned earlier, few women participated in the Vietnam War, but signs of major changes were on the horizon. In 1968, for the first time since the Korean War, the Department of Defense announced plans to increase the number of women in the military services to offset mounting popular opposition to the draft.[46] Four years later, Congress passed the proposed ERA with no mention of the draft, after soundly rejecting the Ervin amendments that would have excluded women from the draft. Later in 1972, the question of drafting women generally ceased to be an issue when President Nixon allowed induction authority to expire and men ceased to be drafted.[47]

With the establishment of the all-volunteer Army in 1973, the question of conscription for either men or women became less pressing. The issue resurfaced, however, in early 1980, when President Carter, responding to the takeover of the American embassy in Tehran and the Soviet invasion of neighboring Afghanistan, made plans to reinstate registration for the draft. Carter was also reacting to increasing criticism of the AVF in the media, which frequently reminded the public that fiscal year 1979 recruiting efforts had fallen short by about 25,000 personnel in the active-duty forces. Carter requested congressional authority to register young women as well as young men for Selective Service, reflecting his strong support for the ERA and other efforts aimed at greater equal-

---

[44] Holm, *Women in the Military*, p. 149; Weigley, *History of the United States Army*, p. 501.
[45] Holm, *Women in the Military*, pp. 157–59.
[46] Holm, *Women in the Military*, pp. 186–87.
[47] Ibid., pp. 186–87, 264; Binkin and Bach, *Women and the Military*, p. 14.

ity of the sexes.[48] Congress, however, was hesitant to address such an emotional issue in an election year. Further, the nation clearly lacked consensus on the appropriate role of women in national defense. In June 1980, the House of Representatives voted to authorize funds for the registration of men only. A report prepared by the Senate Armed Services Committee cited the policy excluding women from combat as "the most important reason for not including women in a registration system."[49]

The constitutionality of the law was tested in the federal courts on numerous occasions, most often as the result of cases brought by men claiming that the exclusion of women from the statute was arbitrary and violated the equal protection clause of the Constitution. Title VII of the Civil Rights Act of 1964 (as amended, PL 88-352, 78 Stat. 24, 2 July 1964) was the primary federal law used to challenge employment discrimination based on race, religion, sex, or national origin. The courts, almost without exception, upheld the constitutionality of the exclusion laws on the grounds that they did not apply to the military and that women could be excluded on the basis of national security needs. Meanwhile, during the 1980s, the issues of women in the military and the draft continued to surface, especially when recruitment goals failed to be met.

Other Gender-Related Issues

In addition to the social, political, and military controversies that served as the backdrop for the Army's efforts to define its structure for training women, a number of other issues remained consistently in the forefront and contributed to the debate about mixed-gender training. The divisions were not always between men and women, nor did the various branches of government or military personnel at any level often agree. The debate was subject to change according to the prevailing military, social, political, and sometimes economic climate.

Physical Strength and Endurance

Concerns about differences between men and women in physical strength and endurance were perhaps the most influential and polarizing in shaping the debate surrounding basic training in mixed-gender units. From 1970 to the late 1990s, the military services, especially the Army,

---

[48] Brian Mitchell, *Women in the Military: Flirting with Disaster* (Washington, D.C.: Regnery Publishing, 1998), pp. 94–95.

[49] Senate committee's report as cited in Holm, *Women in the Military*, p. 363. The president could register men but not actually draft them into the service. Initiating the draft required a separate act of Congress.

conducted numerous tests to determine whether the physiological differences between men and women in upper-body strength, stamina, endurance, speed, and coordination were genetically determined or the product of a less active culture among women and, therefore, subject to change through proper conditioning programs. Test results varied widely except in the case of upper-body strength, which, it was generally agreed, seldom reached the male level among females. Given the importance of upper-body strength for a number of military specialties, especially in the combat arms, these differences had to be taken into consideration in any Army training regimen.

Observers who believed that the physical performance of those entering the military should not be an overriding factor often noted that if the ongoing trend for increased female participation in athletics continued, more women would be capable of performing most physical jobs. Some argued that advanced technology in military equipment and weapons systems would mean a general decrease in the importance of strength and endurance. As one student of the subject put it, "The debate is not (or should not be) over how strong women are, but how strong they need to be."[50] Others pointed to the military's acknowledgment that not even all men were capable of some strenuous combat assignments; thus, it should be recognized that some women *might* be physically capable of meeting the test.[51] Still others believed that no matter how much physical conditioning women undertook, the physical differences in upper-body strength alone made them unsuitable to face the crucible of combat.

Those who believed that compromises regarding physical conditioning and standards would threaten military readiness and combat effectiveness pointed out that body composition and cardiorespiratory factors generally favored men. They maintained that overall size, muscle mass, bone mass, heart and lung size, oxygen intake, body temperature, and sweat-gland function gave men a decided advantage in physical strength, endurance, and heat tolerance.[52] Further, these differences were generally thought to be immutable. Closing with a strong and aggressive enemy (almost sure to be male), engaging in personal combat, and defeating the enemy by the use of one-on-one violence was a highly physical undertaking that modern technology had not changed and was unlikely to change. Given this widely accepted belief, male

---

[50] Stiehm, *Arms and the Enlisted Woman*, p. 219.

[51] Maj Jeffrey N. McNally, "Women in the United States Military: A Contemporary Perspective," unpublished Advanced Research Program paper, Naval War College, Newport, R.I., 1985, pp. 48–49.

[52] Binkin and Bach, *Women and the Military*, p. 82.

trainees, especially new recruits, often thought that integrated physical training for men and women compromised their preparation.

In assessing physical strength and endurance training, especially for new recruits, the services encountered numerous difficulties in setting standards. Should standards be the same for everyone? Most male trainees thought so. Should dual standards be established that judged women for effort rather than accomplishment? Most male trainees thought not. What about "equivalent" training? Some female trainees welcomed concessions to the male-oriented physical training programs while others believed that lower standards did a disservice to women who wished to be judged by the same standards as men. Generally, in the absence of appropriate methods of measurement, many physical standards were unrealistic and neither well defined nor rigorously applied. One Army program introduced in 1982 sought to use physical standards as one criterion for assigning new recruits to MOSs. The program soon lapsed into a counseling tool when the physical-fitness criterion seriously threatened recruitment goals.[53] That experience notwithstanding, establishing physical standards for initial enlistment proved relatively easy compared to the difficulties in establishing and enforcing standards for subsequent service.[54]

Pregnancy, Marriage, and Parenthood

In the continuing discussion about the role of women in the military, the issue of physical fitness standards was the most debated, followed by the issue of pregnancy. As one author observed: "Perhaps the biggest change for military women, though, has been from an implicit agreement to remain childless to the acceptance of an agreement (at least by some women) to remain in service despite pregnancy and motherhood."[55]

Initially, military service for a young woman was thought of as a sort of "temporary career" until she married. No thought was given to the idea that she might become a mother. In fact, in some quarters, doubt remained that a married woman should be allowed to serve in the armed forces.[56] The WAC regarded the pregnancy issue as a threat to its reputation, which was always subject to rumors about the immorality of military women. In truth, wrote WAC historian Lt. Col. Mattie E. Treadwell USA (Ret.), "pregnancy among unmarried [WAC] women

---

[53] This program was known as the Military Enlistment Physical Strength Capacity Test (MEPSCAT).

[54] Stiehm, *Arms and the Enlisted Woman*, p. 193.

[55] Ibid., p. 28.

[56] Prior to World War II, Army nurses were forbidden to marry.

. . . was about one fifth that among women in civilian life."[57] In any case, discharge was mandatory for pregnant WACs and marriage was grounds for requesting discharge. By the 1960s, women were no longer discharged for marriage, and married women could join any of the services. The focus then turned to women with minor children at home, a situation that required a waiver if the woman wished to remain in military service. By the early 1970s, the services were approving up to 86 percent of waivers and losing about 3,000 enlisted women annually to pregnancy and parenthood.[58]

The advent of the AVF in 1973 caused the Department of Defense to reexamine the pregnancy issue. In June 1974, to reduce attrition, the department directed all services to develop new policies by 15 May 1975 that would make separation (honorable discharge) for pregnancy voluntary instead of requiring application for a waiver to remain in the service. The Army requested a waiver from the new policy, which was denied, but implementation by the Army was moved back to November 1975. Meanwhile, the Army continued to try to identify the best approach to dealing with pregnant soldiers. A year later, a study based on the integration of the judgments of field commanders and social science researchers concluded, among other things, that pregnancy resulted in under-strength units and a lack of deployability.[59] Two years later, in March 1978, an Army report concluded: "Unit leaders do not cope well with the entire pregnancy issue. In many cases the women do not pull their share of extra duty, are exempted from field duty, draw full pay and allowances without earning them, and are not required to maintain minimum dress standards."[60]

In yet another study, released in 1982, the Army chose not to address the highly controversial subject and to pass further action on to the Department of Defense. During that year, pregnancy policies changed to allow the services not only to keep pregnant women but to involuntarily retain them if their separations were deemed not in the best interest of the military service.

The picture was not completely negative for pregnant women who chose to remain in the military. Beginning in 1976, the courts usually

---

[57] Treadwell, *The Women's Army Corps*, p. 193.

[58] Stiehm, *Arms and the Enlisted Woman*, pp. 210–13; Holm, *Women in the Military*, pp. 300–303.

[59] Stiehm, *Arms and the Enlisted Woman*, p. 138; U.S. Department of the Army, Office of the Deputy Chief of Staff for Personnel, *Women in the Army Study* (Washington, D.C., 1976), pp. 7–9.

[60] U.S. Army Administration Center, *Evaluation of Women in the Army*, 1978, pp. 1-30, 1-31.

ruled in favor of pregnant servicewomen on the grounds that pregnancy was a "temporary disability" and that involuntary separation was a violation of due process and equal rights.[61] Even arguments that focused on loss of duty time due to pregnancy were undercut by a 1977 Department of Defense study showing that men lost more duty time on average for confinement, desertion, alcohol and drug abuse, and absence without leave than did women.[62]

The services continued to complain that pregnancy and motherhood limited assignment mobility, required special measures to ensure the safety of a woman and her unborn child, and brought the question of day care to the fore, to name only a few issues.[63] Pregnant women could not receive routine immunizations, a policy that precluded overseas deployments. The issue was further complicated by the lack of proof or agreement about the effect of heavy work on the expectant mother and unborn child. By the early 1980s, the focus of discussion had moved from the questions of morality common during the WAC era to issues of readiness and mission accomplishment in the services. Pregnant women were a fact of life in the military; the question was how best to cope with them.[64]

Single parenthood and marriage between service members were also debated in the context of the role of women in military service. Single parenthood for men with custody of children, as well as for women, was perceived by many as potentially affecting readiness for deployment, unaccompanied tours, field exercises, temporary duty assignments, alerts, extended hours, and changes of station.[65] Some commanders argued that single parents should be discharged to assume their family responsibilities. Some senior military women agreed that mothers should not be soldiers. Other military authorities asserted that single women fared better, professionally and financially, in the service than in the private sector.

One of the fastest growing subgroups in the post-AVF were couples in the service, many with children. Military officials worried that in the event of a sudden deployment, both partners might not be willing to go or would leave unattended children behind. Other potential problems included incompatible work schedules and family separation resulting from different duty locations. All the services had programs of joint

---

[61] See *Crawford v Cushman* 531 F2d 1114, 1120 (2d Cir 1976).

[62] Holm, *Women in the Military*, p. 303.

[63] Working with pregnant soldiers was reported to make many males uncomfortable and "uneasy." See Stiehm, *Arms and the Enlisted Woman*, pp. 22, 50.

[64] Mitchell, *Women in the Military*, p. 156.

[65] Stiehm, *Arms and the Enlisted Woman*, p. 221.

assignment, but such assignments became more difficult as the number of military couples increased.[66] Reenlistment rates also suffered. As late as 1989, the Army retained the right to deny enlistment to single parents with custody of children. Service members who became parents after enlistment, however, were not discharged.[67]

These two issues came together during Operations DESERT SHIELD and DESERT STORM in 1990 and 1991, when new questions were raised about the deployment of single parents and dual-service couples with children. According to one author, in 1990, 47,000 dual-service couples with children and 67,000 single parents were in military service.[68] Intense media attention focused on tearful scenes of married women and single parents being separated from their children, generating concern and sympathy among the American public.[69] Partly as a result, a number of bills were introduced in Congress immediately following the Gulf War to prevent the deployment of single custodial parents or the simultaneous deployment of both military parents. The Pentagon position, stated by Assistant Secretary of Defense for Force Management and Personnel Christopher Jehn, was that the same sacrifices had to be made by all military members, married or single, with or without children, as the "only understandable and fair policy and one that is consistent with the American tradition of equality."[70] The debate would continue, but it was clear that "the parenting issue was a natural by-product of the modern all-volunteer force."[71]

Attrition and Retention

In any discussion of the relative conditions of military service for women and men, the issues of attrition and retention and the effects on readiness and cohesion were bound to arise. Concerns about severe loss of personnel, especially among junior enlisted women during their first-term enlistments, predated the all-volunteer Army. Before the Korean War, attrition rates for women and men were similar. At the onset of hostilities in June 1950, a ban was placed on voluntary separation of women who were married. When the ban was lifted a year later, a landslide of

---

[66] Ibid., pp. 217–19.

[67] Mitchell, *Women in the Military*, p. 158.

[68] Holm, *Women in the Military*, p. 465.

[69] Ibid.

[70] House of Representatives, Assistant Secretary of Defense for Force Management and Personnel Christopher Jehn, speaking before the Subcommittee on Military Personnel and Compensation of the House Armed Services Committee, 102d Cong., 1st sess., 19 Feb 1991.

[71] Holm, *Women in the Military*, p. 465.

voluntary discharges occurred until attrition overtook accessions. This phenomenon was exacerbated by the lowering of the minimum legal age in 1948, which had filled the ranks with eighteen-, nineteen-, and twenty-year-olds—then the most popular marriage ages. During the Cold War of the late 1950s and the 1960s, attrition was high among first-term enlisted women as marriage continued to offer a means of escape from the enlistment contract. Of enlisted women, an estimated 70 to 80 percent left military service before completing their first enlistment. That phenomenon meant that 40 to 50 percent of female personnel had to be replaced annually, a rate two and one-half times the replacement rate for men.[72]

Although these high rates of attrition sometimes threatened women's programs and were often cause for concern among military leaders, the draft was still in place, and women made up only approximately 2 percent of total U.S. military strength. That situation changed rapidly with the end of the draft and the advent of the all-volunteer Army. As senior leaders increasingly believed that more women would be necessary to fill the ranks and as the numbers of military women rose to more than 10 percent, high attrition rates became alarming. For those opposed to an increase in the numbers of women in the military or to their inclusion in the services at all, the attrition figures were powerful ammunition. High replacement rates, they asserted, not only reduced service strength but increased "personnel turbulence" and wasted the training investment. Figures for the Army in 1981 showed that 40.3 percent of women failed to complete their enlistment contracts, compared to 23.5 percent of men.[73] A study in the same year by the Army Research Institute for the Behavioral and Social Sciences (ARI) found the figures to be 41 percent for women and 35 percent for men.[74]

Defenders of women serving in the military often argued that women's continued exclusion from combat frustrated their ambitions for promotion and, therefore, contributed to high attrition rates. Further, the problem could be traced not to the caliber of the women the services attracted or their inability to compete—as some had suggested—but to the flawed policies of the male-dominated military services. To opponents of women in combat, these arguments were seen as feeble

---

[72] Holm, *Women in the Military*, pp. 156, 163. The only group with a higher replacement rate consisted of male draftees with two-year tours, which large numbers failed to complete.

[73] Mitchell, *Women in the Military*, p. 151. Figures are taken from Department of Defense, Office of the Assistant Secretary of Defense, *Military Women in the Department of Defense* (Washington, D.C.: Jul 1987), p. 63.

[74] Glenda Y. Nogami, "Fact Sheet: Soldier Gender on First Tour Attrition" (Alexandria, Va.: U.S. Army Research Institute for the Behavioral Sciences, 1981).

excuses and careerist special-pleading rather than substantive reasons focused on mission accomplishment.

The debate continued in all the services. In fact, the high attrition rates for female recruits tended to support those who believed that women in the military hurt readiness, wasted training dollars, and generally detracted from the national defense.

Male Ego, Bonding, and Unit Cohesiveness

The psychology of men and women also entered the debate. War and soldiering, with few exceptions, were almost exclusively male preserves during much of recorded history.[75] What possible damage could be caused to the masculine warrior ethic by the presence of women in the military? For those opposed to mixing the sexes in the military, commitment to the group and strong unit cohesion were seen as dependent on male bonding. In short, they agreed, the presence of women in what were previously all-male organizations and situations had the potential to seriously disrupt men's interpersonal relationships and, thus, their unit effectiveness.[76]

The question of women in formerly all-male military units was a deep concern for military leadership. Although many maintained that with advancing technology, almost anyone could learn military skills, some believed that unit effectiveness and cohesion were far more the result of "sociopsychological" bonding among soldiers: "Without this crucial bonding, units disintegrate under stress no matter how technically proficient or well-equipped they are. The key variable in the effectiveness of a military unit is not the technical abilities of its troops . . . but the ability of troops to maintain cohesive bonding groups under fire. . . . We have tinkered with the very foundations of our military forces without any sound sociological or psychological research from which to predict the results of our organizational restructuring."[77]

One observer of Army basic training wrote, "All-male companies regularly exceeded training standards for tests of motivation and endurance, such as the twelve-mile road march, while integrated companies rarely exceeded standards for such events."[78] Another commentator on

---

[75] Jeff M. Tuten, "The Argument against Female Combatants," in *Female Soldiers: Combatants or Noncombatants: Historical and Contemporary Perspectives,* ed. Nancy Loring Goldman (Westport, Conn.: Greenwood Press, 1982), p. 239.

[76] Mady Wechsler Segal, "The Argument for Female Combatants," in *Female Soldiers,* ed. Nancy Loring Goldman, pp. 228–29; Tuten, "The Argument Against Female Combatants," p. 239.

[77] Richard A. Gabriel, "Women in Combat: Two Views," *Army,* March 1980, 44.

[78] Mitchell, *Women in the Military,* p. 175. Brian Mitchell served on the 1993 Presidential Commission on the Assignment of Women in the Armed Forces.

the idea of mixed-gender combat units suggested that male objections to women in combat or combat training were based on the premise that the male ego demanded that men remain the protectors of women and women, consequently, the protected.[79] What would be the effect on cohesion and esprit de corps in traditionally all-male units when the sex roles of warrior and protector were blurred?

Proponents of gender-mixed units and training programs generally believed that women—either from a duty or opportunity standpoint—should participate in the military fully as equals. Some who held the strongest views maintained that when women were able to perform many traditionally male jobs successfully, military service would cease to validate manhood or masculinity.[80] It would be a risky leap into the unknown. Others pointed out that arguments over the effects on unit cohesion were reminiscent of those used in the past to justify excluding women from such occupations as law, medicine, law enforcement, and firefighting. Further, arguments used against women's participation were similar to those used against African American men serving in racially mixed units or in combat. Proponents insisted that throughout Operation DESERT STORM, "When the action started, the mixed units and crews bonded into cohesive, effective teams."[81]

Opponents and proponents of mixed units appear to agree on only one point: Despite dozens of studies, little to no evidence exists to support either position. In the end, most observers adhered either to the premise that the first imperative for the armed forces was the highest possible level of combat readiness or to the opposite view that social justice and equity as national objectives were equal or superior to the cause of military preparedness. Lacking a reasoned middle ground, the Army chose not to address the issue in the important and influential 1981–1982 report titled *Women in the Army Policy Review*.[82]

Emotions and Aggression

Most of the literature on the differences—or lack thereof—between men and women with regard to emotions and aggression is focused on

---

[79] Larry B. Berrong, "A Case for Women in Combat," U.S. Army Command and General Staff College, Jun 1977, p. 3.

[80] Helen Rogan, *Mixed Company: Women in the Modern Army* (New York: Putnam, 1981), p. 91.

[81] Holm, *Women in the Military*, p. 463.

[82] Department of the Army, Office of the Deputy Chief of Staff for Personnel, *Women in the Army Policy Review* (Washington, D.C.: 12 Nov 1982).

the pros and cons of women in combat. The arguments surrounding this question, however, nearly always arose when the topic under discussion was mixed-gender training companies. A substantial percentage of those involved in the decision-making process concerning mixed Army BCT units, as well as many in the public sector and in Congress, believed that the establishment of mixed units at the basic-training level would inevitably lead to women's inclusion in the combat arms and participation in combat in a future conflict.

Central to the debate was the question of whether significant differences exist in the way men and women behave. If so, what were the implications for mixed-gender units? Some who favored mixed units argued that differences in the nature of men and women were the result of social conditioning and stereotyping, not heredity or biology. Thus, a change in the socialization process could, over time, produce women with the same aggressiveness and fighting spirit as men. Other students of the subject saw profound temperamental differences between the sexes and asked how these differences affected behavior and effectiveness in situations of war and peace, life and death on the battlefield. Some argued that modern warriors did not need to be as aggressive or strong as in the past. Indeed, certain acknowledged inherent differences between the sexes, such as the idea that women were generally better suited to performing routine and intricate tasks than men, might offer an advantage in modern warfare. Further, some questioned whether the most aggressive individuals were likely to be the most effective and disciplined soldiers. Regarding control of emotions amid the fear and stress of battle, little empirical evidence existed to confirm or refute the notion that women were less capable of self-control in violent situations; those conditions had simply never been tested. One female member of one of the first mixed-gender training units, drawing on the words of noted anthropologist and writer Margaret Mead, wrote: "[Because women] have grown up with little sense of the ritual nature of war, they would fight ruthlessly and without scruple, as they always have, fighting like the underdog, paying scant attention to the chivalric rules."[83] This view, however, was mere speculation (or perhaps even wishful thinking), given that women had not been tested in the ultimate laboratory—as part of teams engaged in sustained attempts to close with enemy soldiers and kill them.

Opponents to the participation of women in the military or in combat were equally convinced that women lacked the "killer instinct" for such employment. The editor of one study wrote: "For many Americans,

---

[83] Rogan, *Mixed Company*, p. 283.

the battlefield remains a unique workplace, where soldiers are required not only to be physically strong and emotionally aggressive, but also brutal and capable of killing. Many Americans are still unprepared to acknowledge these qualities in women."[84] Nearly all those who opposed mixed units believed that women were innately less combative than men. For some, the most important concern was that introducing women into units, especially combat units, would further confuse an already confusing environment and have a negative effect on male fighting performance.

Thus, the battle lines were drawn; on one side were those concerned that women in the military would compromise readiness, and on the other, those who believed that women would not have full equality until they served along with men in all environments and situations.

Fraternization

One analyst of shifting policies toward women in the military wrote: "The most troublesome issue was that of fraternization between officers and enlisted personnel of the opposite sex."[85] The increase in numbers of females in the military, and the increasingly mixed-gender nature of units, caused that issue to spill over into all levels of the military hierarchy. To maintain impartiality, discipline, and morale, military policy had traditionally prohibited close personal relations and social activities between officers and enlisted personnel or seniors and their subordinates. Such "fraternization" would have an extra twist when both sexes were involved. Concerns about relationships between males and females and superiors and subordinates, along with the use of power and seduction, were sure to arise in the training of young men and women in mixed units.

Enforcement of anti-fraternization rules was uneven across the military services and even within each service. Much of the enforcement relied on custom and tradition. The greatest disdain existed for female officers dating enlisted men; less harshly judged but still strongly discouraged were relationships between male officers and enlisted women. Although obliged to support it publicly, the WAC director during World War II, Col. Oveta Culp Hobby, never agreed with the rule regarding male officers and enlisted women, nor did Army General Dwight D. Eisenhower and General George C. Marshall. However, the Army chief

---

[84] Carol Wekesser and Matthew Polesetsky, eds., *Women in the Military: Current Controversies* (San Diego, Calif.: Greenhaven Press, 1991), p. 13.

[85] Holm, *Women in the Military*, p. 73.

of personnel declared, "The traditional relationship between officers and enlisted personnel is a strongly entrenched custom of the service, and any exception which is made for WACs will be a step in the direction of its complete elimination."[86] After the war, the issue generally disappeared, only to surface again when WACs were integrated into the regular Army. Although most experts generally agreed that a joint policy was needed, nothing was done, primarily because the issue was a public relations liability.

The Army, in general, left fraternization issues to local commanders until late November 1978, when the service published Army Regulation 600-20, *Army Command Policy.* The Army version differed from the traditional understanding in that it applied only to those in the direct chain of command. The regulation was revised in November 1984 to include examples of relationships that could lead to the perception of partiality or undermine morale, discipline, or authority. The regulation and its subsequent versions were adopted primarily to prevent dating between trainers and trainees during BCT. Within the regulation itself was the caveat "no canned solutions are available."[87] Disciplinary authority for suspected instances of fraternization resided with commanders according to the Uniform Code of Military Justice.[88] Commanders across the Army complained that the regulation was too subjective and set no clear rules for what was proper or posed a threat to morale.

Despite its lack of specific guidance, the 1984 regulation remained in effect—with several updates—until March 1999.[89] In 1997, shortly after allegations of violations of fraternization policies at several Army installations, Secretary of Defense William S. Cohen established a joint task force to examine whether current policies and practices for maintaining good order and discipline across the armed services were fair and effective. Subsequently, a Department of Defense directive of 29 July 1998 tasked all the services to align their fraternization policies to "establish uniform service policies and regulations governing fraternization." Because Army policy was perceived to be more liberal than that of the other services, the changes for that service were, in turn, more extensive and restrictive.

---

[86] Ibid., p. 75.

[87] Stiehm, *Arms and the Enlisted Woman*, p. 209.

[88] The Uniform Code of Military Justice was adopted by the U.S. Congress effective 31 May 1951 to standardize the rules of behavior of the services.

[89] AR 600-20, *Army Command Policy*, was updated in Aug 1986 and combined with AR 600-21 and AR 600-60 in Mar 1988. The current version of AR 600-20 became effective 13 Jun 2002.

On 2 March 1999, the Army announced the revised "good order and military discipline" policy. The greatest change from earlier policies was the prohibition of *any* relationship between soldiers of different ranks if the relationship appeared to compromise supervisory authority or could result in preferential treatment. Officers and enlisted men who were currently dating were given until 1 March 2000 to marry or end their relationships. Relationships between members of the training cadre and initial-entry trainees were expressly forbidden. Soldiers of different ranks were prohibited from engaging in private business deals or long-term business relationships, gambling, borrowing, providing tobacco products, or attending events other than Morale, Welfare, and Recreation activities. The new policy also applied to relationships of Army personnel with service members from other branches of the military. The new fraternization policies went into effect immediately; however, the new regulation, AR 600-20, *Army Command Policy,* was not approved until 15 July 1999.[90]

Sexual Harassment

As the number of women in the Army grew from 2 percent in the early 1970s to a figure approaching 15 percent in the opening years of the twenty-first century, sexual harassment caused increasing concern for military leadership, soldiers, officers, and some public focus groups. A special concern was the perceived vulnerability of young enlisted women in basic training programs. A large part of the problem—for all the military services—was defining what constituted sexual harassment or, indeed, harassment of any sort. If demands by superiors for sexual favors clearly constituted harassment, then what about lewd remarks, verbal abuse, casual touches, or consensual sexual relationships? At what point did tough and demanding training become abuse? Were *sexual harassment* and *sexual misconduct* synonymous terms?

Although Army senior leaders were aware that the increase in percentages of women in the ranks might encourage more incidents of sexual harassment, the service did not address the issue until 1980. At that time, a congressional committee released a report entitled *Sexual Harassment in the Federal Government,* which gave visibility and, perhaps, legitimacy to the subject.[91] Concurrently, the Army made public

---

[90] Army News Service, Department of the Army Public Affairs, 30 Jul 1998, 2 Mar 1999; AR 600-20, 26 Jul 1999.

[91] U.S. House of Representatives, Subcommittee on Investigations of the Committee on Post Office and Civil Service, 96th Cong., 2d sess., 30 Apr 1980.

a survey suggesting that sexual harassment was definitely a problem and publicly court-martialed two male soldiers for using indecent language to a female soldier.[92] The incident brought on-post investigations by the Army inspector general and more congressional hearings. The Army was quick to develop enforcement policies to aid commanders in ending harassment as a part of their duties. Over the next two years, the Army also launched a number of educational programs, including several for initial-entry personnel. Although incidents continued to some extent, from 1980 to 1996, no highly visible cases or public outcry were seen.

Then, in November 1996, the Army announced that it was charging five soldiers assigned to the Ordnance School at Aberdeen Proving Ground in Maryland with a variety of sex crimes. The arrests, coming so quickly after the Navy's experience with the Tailhook scandal in September of that year, precipitated more training in the prevention of sexual harassment, the establishment of a sexual harassment hotline to report incidents of harassment, and congressional hearings. Prosecutions spread to other Army installations, and ultimately, a dozen drill instructors were charged. The Secretary of the Army quickly appointed a Senior Review Panel on Sexual Harassment with active and retired male and female service members and civilians to study the issue.[93] In June 1997, the Department of Defense established the Federal Advisory Committee on Gender-Integrated Training and Related Issues to provide its own review of military training.

Meanwhile, the Army began to revise Army Regulation 600-20, *Army Command Policy*, to substantially strengthen its sexual harassment policy. Even as the new regulation was initially fielded on 15 July 1999, Congress mandated the formation of another panel, the Commission on Military Training and Gender-Related Issues, known as the Blair Commission, to further study the training of men and women in mixed-gender units. These panels and commissions addressed a number of issues from numerous perspectives and elicited a wide variety of responses to their reports. Despite the differences of opinion, the results of these study groups and of others before them indicated that sexual harassment continued to defy adequate definition.

---

[92] Stiehm, *Arms and the Enlisted Woman*, pp. 205–06.

[93] Mitchell, *Women in the Military*, pp. 309–11. The school at Aberdeen conducts advanced individual training (AIT), not basic combat training (BCT). Only four of the drill sergeants charged were sent to prison. The commandant and three other senior officers received letters of reprimand.

Public Opinion

In 1977, the authors of a Brookings Institution study on women in the military stated: "Women's role in the armed forces will ultimately depend on the extent to which national institutions—social, political, judicial, and military—are willing to break with their past—a past reflecting a persistent pattern of male dominance."[94]

Indeed, it was clear that women's roles in the armed forces would depend, in large part, on the public's view of the issues. Public opinion had far-reaching implications for the attitudes and actions of Congress, the courts, and military leadership. But the opinions of the American people were difficult to pin down. Results of national opinion polls and surveys on the subjects of the draft, women in combat, mixed-gender training, and job assignments, to mention only a few, tended to vary widely. Some patterns, however, were discernible. In general, the gradual acceptance of opportunities for women in the civilian sector tended to be reflected in the gradual acceptance of an increased role for women in the military. For example, in a poll conducted in 1971 to determine opinions about whether women should have equal treatment regarding the draft, 71 percent disagreed with the idea.[95] A similar survey in 1982 indicated that of those who favored a return to the draft, slightly more than 50 percent favored drafting women.[96] Change in public attitudes was slow but not at a standstill.

National attitudes toward women in the armed forces were certainly ambivalent. Some of those who responded to the survey justified restricted opportunities for women on the grounds that they might become prisoners of war or be killed. Others, in almost equal numbers, maintained that equal service in the military was a necessary component of full citizenship. Measures of the public will were also conveyed through elected officials. Although attitudes in Congress were not sharply drawn, there appeared to be little opposition, at least in principle, to women in the military until the discussion focused on women in combat. In 1975, however, Congress approved the proposed ERA for ratification and the Stratton Amendment to Title 10 of the U.S. Code (PL 94-106), which allowed women to attend the military academies.[97] Congress did

---

[94] Binkin and Bach, *Women and the Military*, p. 39.

[95] Ibid., p. 39. The poll was conducted by the Roper Center for Public Opinion Research at the University of Connecticut.

[96] Stiehm, *Arms and the Enlisted Woman*, p. 182. The poll was conducted by the National Opinion Research Center at the University of Chicago.

[97] The House of Representatives vote on the change to Title 10 was 303 to 96. The Senate approved by voice vote.

not seem to favor or oppose the various measures as much as it resisted publicly debating them.

These attitudes, preconceptions, political goals, and social agendas came into conflict during Army debates on women in the service from the 1970s through the 1990s. Nowhere did these arguments play out more fully than in the tentative and haphazard approach of the Army to mixing men and women in the units responsible for the initial training and socialization that turned civilians into soldiers: the basic training companies.

II

The First Experiment
Basic Combat Training, 1975–1982

● ●

By the mid-1970s, the necessity to develop a program for training males and females together in basic combat training (BCT) was clear to the Army leadership.[1] A detailed account of the Army's experience must answer at least three central questions: Why did the U.S. Army undertake that effort at that particular time? What was the trainees' experience with the first mixed-gender companies? Why, after only about three years, did the Army abandon the experiment and again segregate basic training units?

The Experiment

The Army's decision regarding mixed-gender basic training was the result of a number of social currents—some general, some specific—that came together in the 1970s to generate pressure on the service to re-examine its basic training policies. Beginning in the late 1960s, the women's liberation movement brought a greatly enhanced awareness of women's rights issues, especially in the workplace. The most visible symbol of equal rights and full citizenship for women, the Equal Rights Amendment (ERA), began to make its way to state legislatures in 1972. There appeared, at that time, to be little doubt of its approval.[2] Consequently, the Army leadership believed that the amendment's adoption would seriously compromise their ability to determine policy for women in the Army. In short, many senior officials believed it important to establish policy before it was determined elsewhere. Thus, time seemed to be of the essence. Those factors, plus a substantial

---

[1] Men and women already trained together in some advanced individual training (AIT) programs by the mid-1970s.

[2] RMC Roxine C. Hart, "Women in Combat," February 1991, p. 14.

decline in recruitment, led the leadership to look for new enticements for women to enlist, such as coed training at the entry level, and to begin to consider how to train women and to establish policy about what roles women could assume in the force.

Army policy regarding BCT was affected by a number of factors. The Navy and Air Force had, for a number of years, trained the sexes together as recruits. In addition, at the officer level, the Air Force had allowed women in the Reserve Officers Training Corps in 1969 on a test basis, followed by the Navy and the Army in 1972.[3] Then, in 1973, conscription ended and the Army became an all-volunteer force (AVF), raising deep concern about the ability of the service to attract a sufficient number of men to fill recruitment quotas and questions about what the policies toward women should be, especially if their numbers increased dramatically. One of the most telling signs of rapid changes taking place in the defense arena was the decision in 1975 to allow women in the service academies for the first time during the 1976–1977 academic year.[4] In addition, in 1978, in response to a court order, Congress voted to allow women to serve on noncombat and combat ships for a period not to exceed 180 days.[5]

As these events unfolded, questions about the future of the Women's Army Corps (WAC) increased. In March 1973, the secretary of the Army directed Army staff to draft legislation to eliminate the legal requirement for a separate WAC. That effort was temporarily put on hold pending other actions. However, beginning in mid-1974, various elements of the WAC disappeared quietly from Army installations. In early October 1974, the death of Army Chief of Staff General Creighton Abrams took away one of the WAC's strongest supporters. In June 1975, the secretary of the Army told Congress that the WAC was no longer needed as a separate corps and that the corps' disestablishment would fully integrate women into the Army. Congress agreed. Finally, in October 1978, President Jimmy Carter signed a bill into law abolishing the WAC and fully integrating its members into the regular Army.[6]

---

[3] Holm, pp. 181–83, 267–73.

[4] The gender integration of the military academies was effected by the Stratton Amendment to Title 10 of the U.S. Code (PL 94-106), which President Gerald Ford signed into law.

[5] Wekesser and Polesetsky, p. 40.

[6] For a full account of the disestablishment of the WAC, see Lt Col Bettie Morden USA (Ret.), *The Women's Army Corps, 1945–1978* (Washington, D.C.: U.S. Army Center of Military History, Army Historical Series, 1990), pp. 310–18, 395–97.

Studies and Experiments: The 1970s

With the WAC disestablished, the question of training arrangements, which had been debated for several years, became even more crucial. Concurrently, a change in the presidential administration seemed to allow no more hesitation. In the 1976 election, Democrat Jimmy Carter defeated incumbent Republican President Gerald Ford. The Carter administration fully supported the women's rights movement, the ERA, and equality in every area of government and society. Although study of the issue predated Carter's term in office, it would have been politically difficult for the Army not to experiment with mixed-gender basic training, although much of the leadership harbored strong misgivings.

In considering how the new program for BCT should be structured, the Army turned to a time-honored strategy: multiple studies. Beginning in late 1974, numerous studies, large and small, focused on the numbers of women the Army needed and on "utilization," because assignment to a military occupational specialty (MOS) determined combat duty. A number of these studies are worthy of attention here, because information gathered during their conduct and the implementation of their recommendations would be used and cited in the debate about training men and women together in BCT. For example, to observers inside and outside the Army, training in mixed-gender units at the entry level seemed naturally to lead to women in combat MOSs. Similarly, the greater the percentage of women in the Army, the greater the concern in some quarters regarding integrated units.

Even before the end of the draft, personnel of the Army's Office of the Deputy Chief of Staff for Personnel (DCSPER) had expressed keen interest in an expansion of the WAC to meet what seemed sure to be a "manpower gap." The establishment of a WAC expansion steering committee institutionalized the expansion effort, which was also supported by the WAC leadership. The committee's efforts resulted in a steadily increasing number of WAC service members. This success, in turn, led to other concerns. In November 1974, the assistant secretary of the Army for Manpower and Reserve Affairs asked the committee to consider the effects that a higher percentage of women would have on the Army and its combat readiness. He wrote in a memorandum to DCSPER: "We do not have a clear answer to the question 'how many women do we want in the Army—unit by unit, MOS by MOS?'. . . Until we can nail this down, we may be setting objectives that are meaningless."[7]

---

[7] Memo, ASA (M&RA) to DCSPER, 7 Nov 1974, no sub, Center of Military History (CMH), as quoted by Morden, p. 369.

Consequently, DCSPER directed the commander of the Military Personnel Center (MILPERCEN) to develop computer models to determine how many enlisted women the Army needed, MOS by MOS. Concurrently, he asked Gen. William E. DePuy, commander of the new Training and Doctrine Command (TRADOC), to analyze the Army's manpower requirements to determine how many enlisted women a TOE (Tables of Organization and Equipment) unit could hold "without degrading its ability to perform its mission."[8]

The MILPERCEN models predicted that up to 8 percent of the Army's enlisted spaces and 16 percent of the officer spaces could be filled by women. The TRADOC report, which was completed in April 1975 and titled *WAC Content in TOE Units*, established the percentage of women (0 to 50 percent) that could be assigned to combat support and combat service support units based on how close to the combat zone these units usually operated. The closer a unit's operations were to the battlefield, the lower the percentage of women who would be assigned to the unit, down to zero. In forwarding the report, DePuy expressed doubts about its utility: "There is no perfect way to arrive at a maximum ceiling on the number of women who can be assigned to TOE units. . . . [This] is largely a subjective exercise."[9] However, in evaluating the report, most major commanders found the TRADOC percentages acceptable. The DCSPER then asked MILPERCEN to subject the TRADOC percentages to its computer model to determine the requirements for enlisted women by MOS and by unit. When the resulting requirement figures were 40,000 more than the Department of Defense expansion goal, the DCSPER decided to conduct a field test of the TRADOC formula before making the results public.[10]

During the next year and a half, the U.S. Army Research Institute for the Behavioral and Social Sciences (ARI) and the Army Forces Command (FORSCOM) designed a test for the TRADOC requirements figures, titled "Women Content in Units Force Development Test" and known as MAX WAC. The field experiment was designed to test the effect of different percentages (10 to 35) on the operational capability of company-size support units. In October, 1976, forty companies—eight each of medical, signal, military police, transportation, and maintenance—at nineteen posts in the continental United States and Hawaii took part in a three-day field exercise. A team of officer

---

[8] Ltr, DCSPER to Depuy, 13 Dec 1974, sub: *WAC Content in TOE Units*, CMH, as quoted by Morden, p. 370.

[9] Ltr, Cdr TRADOC to DCSPER, 9 Apr 1975, sub: *WAC Content in TOE Units*, CMH, as quoted by Morden, p. 370.

[10] Morden, p. 371.

observers conducted performance tests, and ARI and FORSCOM representatives collected interviews with company personnel to evaluate how well each unit performed its mission.[11] In the spring of 1977, the tests were repeated in the same units using a different percentage of women. The findings reported in October 1977 indicated that the content of up to 35 percent of women in the unit had far less effect (5 percent) on unit performance than did such factors as leadership, training, morale, and personnel turnover. Of importance to this study of BCT, the study group concluded that women were not receiving adequate basic training.[12]

The test results disappointed many of the senior leadership, none more so than Maj. Gen. Julius W. Becton Jr., commander of the U.S. Army Operational Test and Evaluation Agency (OTEA), who later became TRADOC's first deputy commanding general with responsibility for all initial-entry training (IET) in the Army.[13] Becton apparently believed the percentages were too high, especially given the wide gap in the quality of training between men and women. An OTEA critique of the preliminary findings of the ARI study concluded that the methodology was faulty and that the data did not support the findings. Becton made several recommendations, among them the gender integration of basic training. When the final study results confirmed the initial findings, the Army chief of staff directed Becton to analyze the design, methodology, and findings of the test. Becton's study showed that the units involved could support no more than 20 percent of women, and he recommended that women's performance be measured in a much longer field test under more realistic conditions.[14]

In January 1976, while planning for the MAX WAC study was still under way, the DCSPER directed his staff to undertake yet another study to examine the many changes that had taken place so rapidly in the Army and to address the expanded utilization of women. He was concerned about pressure from the Office of the Secretary of Defense for even greater increases in the number of women to offset a large deficit

---

[11] Performance assessment was conducted according to the recently developed Army Training and Evaluation Program, which replaced the older Army Training Program and the Army Training Tests.

[12] "Women Content in Units: Force Development Test (MAX WAC)," 3 Oct 1977, Alexandria, Va.: U.S. Army Research Institute for the Behavioral and Social Science (ARI), pp. I–2, IV–29; Morden, p. 371; Stiehm, pp. 140–41.

[13] Following his tenure with the Army Operational Test and Evaluation Agency (OTEA), Becton served as commanding general, VII Corps, U.S. Army, Europe. Becton was TRADOC commander from 21 Jul 1981 to 26 Aug 1983.

[14] OTEA Review and Evaluation of MAX WAC Study, Jul 1977, appendix to MAX WAC report of 3 Oct 1977.

of enlisted men. The study, published in December 1976 and known as *Women in the Army* (WITA), sought to integrate information gained from questionnaires completed by major commanders, an examination of policies and procedures concerning women, and evidence drawn from social science research about women. Of special interest here was the inclusion of BCT as one of the areas to be investigated. Generally, the commanders reported that neither pregnancy nor single parenthood had proved a significant problem. They recommended that women be allowed as far forward in a combat zone as necessary to perform non-combat duties. According to the report, the greatest concern, as usual, was women's relative lack of physical strength, which appeared to be the major differentiating factor between the performance of men and women. The commanders also believed women needed more physical, weapons, tactical, and field training at the BCT level.[15]

Generally, the WITA study called for more research. The issues of pregnancy and single parenting would have to await more data before changes in policy could be considered. Likewise, more research was needed to determine physiological, psychological, and sociological effects on women in nontraditional roles and their reaction to combat. In one of its most controversial moves, and one that might affect BCT, the study group recommended that six combat support MOSs be closed to women and that thirteen others be temporarily closed until some career programming problems could be resolved. Further, the study recommended that additional studies dealing with actual physical training be completed to determine the rate of attrition due to physical incapacity. Regarding basic training, WITA noted that "women do not receive the same basic entry training [as men]" and announced that TRADOC would test a common-core basic initial-entry training (BIET) program at Fort Jackson, S.C., from September to December 1976, using both male and female recruits.[16] Throughout the study, the focus was on "the impact of military women on the ability of the armed forces to carry out their missions" and the report continued: "The basic premise on which Army policy is founded—exclusion of women from direct combat roles—is a sound one....It is clear that the original intent of Congress

---

[15] For a discussion of TRADOC commander Gen William E. DePuy's comments and suggestions regarding the WITA study, see TRADOC Annual Historical Review, FY 1976, pp. 315–16 (CONFIDENTIAL—Information used is UNCLASSIFIED). Hereafter cited as TRADOC AHR with the fiscal year.

[16] *Women in the Army Study*, Office of the Deputy Chief of Staff for Personnel, headquarters, Department of the Army, Dec 1976, Executive Summary. Hereinafter cited as WITA, 1976.

and, by extension, the intent of the American people, was that women perform in noncombatant roles."[17]

The WITA study concluded, "while there is considerable work left to do, the Army is on the right track. The current plan for women is acceptable and will not lead to an organization that will be ineffective in time of war."[18]

In the spring and early summer of 1977, the Department of Defense and the Army published studies regarding how women should be used in the military given their ever-increasing numbers. Underlying these studies is a thinly veiled concern about the compromise of combat effectiveness and readiness. Although these studies did not directly affect the dynamics of mixed-gender training at the basic level, they did affect what MOSs female recruits could choose and any changes in the program of instruction. The DoD study, titled *Use of Women in the Military*, begun in January 1977 and published in May, concluded that "continued expansion of the number of enlisted women used in the military can be an important factor in making the all-volunteer force continue to work."[19] Generally, the Defense Department's study, as with all the other studies conducted in the late 1970s, cited a need for more research and additional studies before specific conclusions could be reached.

An important feature of the DoD report of 1977 was a submission by each of the services outlining plans for doubling the number of enlisted women by 1982. Only the Marine Corps agreed to the increase. The Navy and Air Force maintained that problems in managing rotations meant that they could not implement such an increase until Congress removed restrictions on women serving aboard ships and airplanes. The assistant secretary of Defense accepted the data. The Army's objections were not received as easily. The assistant secretary directed that the Army develop a plan to increase enlisted strength gradually but substantially by 1982. Even before publication of the DoD study, the Army leadership was informed that according to plan, the numbers of enlisted women would rise from the projected 50,400 to 100,000 in 1983. Immediately, the Army DCSPER established a task force under the direction of MILPERCEN to evaluate the impact of such an expansion. The report of that task force became the *Utilization of Women in the Army* study of 30 June 1977.

---

[17] Ibid., pp. 1–6.
[18] Ibid., pp. 1–7.
[19] "Use of Women in the Military," Office of the Assistant Secretary of Defense for Manpower, Reserve Affairs, and Logistics, Washington, D.C., May 1977, p. iv.

The MILPERCEN task force, using an extensive data collection plan, examined information on recruiting, training, assignment, promotion, deployability, and unit readiness. The task force concluded that the Army could meet the goal of 100,000 by 1983, but that such a rapid expansion would severely affect promotion, distribution, assignment, and rotation matters. Other concerns included the reluctance of women to enter nontraditional fields and a "disproportional" number of promotions for women if large numbers were quickly recruited at the entry level. The study group's final recommendation was that the Army should not attempt to meet any specific force level for women. The service did agree, however, to abandon the policy of giving women "leftover" training seats and allow them to reserve a time and a specialty for enlistment, as men were allowed to do.[20]

Almost as soon as the Army's MILPERCEN study reached his desk, the DCSPER directed still another study to determine how many women, by MOS and grade, the Army could absorb without compromising the accomplishment of its worldwide ground combat mission. The Army chief of staff assigned the study, known as the "Evaluation of Women in the Army," or EWITA, to begin in August 1977, to the commander of the Army Administrative Center in Fort Benjamin Harrison, Ind.[21] At about the same time, the DCSPER requested that TRADOC "evaluate the combat role from which women should be excluded and propose a definition for this exclusion." TRADOC's action in that regard would support HQDA, which was developing a new policy on combat exclusion for women on the grounds that Army policy for assigning women was too restrictive and precluded full utilization of women. A definition of the combat role from which women would be excluded was proposed: "Women will be excluded from positions that have the primary function of engaging in sustained combat in units with the primary mission of closing with and destroying the enemy, or seizing and holding ground."

The same caveat would apply also to certain positions in support units. Beyond that, TRADOC referred any further action on the employment of women to the Administrative Center and its EWITA study.[22]

The voluminous EWITA report was published in May 1978. It was based on questionnaires completed by TRADOC's installations and staff offices, interviews with commanders, and other data. Apparently

---

[20] Morden, pp. 375–78; Stiehm, p. 140.

[21] "Evaluation of Women in the Army" (EWITA) study, Executive Summary, p. 1-1.

[22] Decision paper with attachments, DCSPER to TRADOC chief of staff, Jul 1977.

ignoring TRADOC's proposed definition of combat exclusion, the team adopted a policy directing that "women will be excluded from positions which have as their primary function the crewing or operation of direct and indirect fire weapons" (unless a unit's weaponry could be fired only in a non-line-of-sight mode).[23] Further, the study argued that theoretically, the Army could support 159,700 enlisted female positions but at the cost of recruiting women with lower intellectual, educational, and physical qualifications. Regarding the ongoing physical strength issue, the study group recommended that the DCSPER establish "gender-free" standards for each MOS and design physical training for recruits to enhance MOS capability. Other recommendations included the opening of three MOSs to women and the closing of fifty-one MOSs for reasons of the combat ban, career progression, or physical requirements.[24] The EWITA study was, predictably, controversial, with the greatest dissension coming from within the Army. Of forty-nine recommendations, only fourteen were approved for implementation.[25]

Concurrent with the EWITA study, ARI conducted the follow-on to the MAX WAC study of 1976 that Maj. Gen. Becton of OTEA had suggested. This time, the TRADOC formula for the recommended percentages of females in units would be tested for ten days during the annual REFORGER (Return of Forces to Germany) field training exercises. A total of 229 women accompanied the REFORGER 77 troops to West Germany, where observer teams evaluated the women's performance and ability to adapt to field conditions.[26] The ARI report, titled *Women Content in the Army—REFORGER 77* (REF WAC 77), published in May 1978, concluded that "the presence of female soldiers...did not impair the performance of combat and combat support units."[27] The team found that women were as proficient as men in most MOSs. But, eighteen of ninety MOSs were found by 50 percent or more of the supervisors to be too physically demanding for women. The REFORGER 77 report also cited leadership and management problems, problems with weapons and tactics training for women, and difficulties adapting to billeting in the field. Throughout the exercise, according to the final report, some men had difficulty accepting women's participation. One

---

[23] EWITA, pp. 1–3.

[24] Ibid., pp. 1-19–1-31.

[25] Morden, p. 381.

[26] Msg, Department of the Army headquarters to distr, 301309Z Jun 1977, sub: *Women Content in the Army–REFORGER 77.*

[27] *Women Content in the Army–REFORGER 77*, ARI Special Report S-7, 30 May 1978, pp. 1–2.

sergeant was quoted as saying that he "simply did not want women in 'his' Army."[28]

With the publication of the REFORGER 77 report, the Army had three studies concerning the optimum percentage of women in units, all with different conclusions. It also had a number of studies that looked at the MOSs to which women should be assigned, especially in view of the combat ban. These studies, too, supplied a variety of answers. Collectively, the studies of the late 1970s represented the Army's reaction to the military, social, and political issues discussed in Chapter I. Generally, the Department of Defense was more supportive of expanded opportunities for women in the services than was the Army itself. A partial explanation appears to have been a belief that because the Navy and the Air Force were encountering only minor difficulties with the integration of women, the Army's reluctance was somewhat arbitrary. The Army leadership, on the other hand, cited the multifunctional nature of the Army's ground combat mission and reasoned that it had already taken the brunt of the expansion in numbers and percentages since 1973. The varied results and interpretations of the studies were used—and continued to be used—to support pre-existing viewpoints.[29] These and subsequent studies, by subject and substance, helped inform the Army's and TRADOC's decisions to establish a mixed-gender basic training program.

The Fort Jackson Test

As the studies continued, the Army leadership, after June 1975, believing that the WAC would be disestablished and that the proposed ERA would be approved, began to plan for training the former WACs in integrated units. As a result, the Army's TRADOC began to make and evaluate a number of changes in the basic training program of instruction (POI). The command also made plans to test integrated training during BCT at Fort Jackson.

As fields previously closed to women gradually opened and women began to move into nontraditional military jobs, it became clear that knowledge of weapons and defensive tactics would be necessary. On 26 March 1975, on the recommendation of DCSPER and TRADOC commander General William E. DePuy, those subjects became mandatory in WAC training for those enlisting or re-enlisting after 30 June 1975. The

---

[28] Ibid., pp. IV 7–11, quotation p. IV 9; Morden, p. 372; Stiehm, p. 142.

[29] See, for example, the various interpretations of the REF WAC 77 report in Morden, p. 372; Stiehm, p. 142; Mitchell, *Disaster*, p. 84; and Holm, pp. 257–58.

WAC training staff revised the basic training program to include qualification on the M-16 rifle beginning in December 1976, and defensive
techniques, such as digging foxholes. In July 1976, TRADOC added
training on weapons, such as light antitank weapons, grenade launch-
ers, Claymore mines, and the M-60 machine gun, for enlisted women at
Fort McClellan, Ala. In the spring of 1977, training with hand grenades
was also added.

Physical training was increased to include more exercise, and the
day march was expanded from two and a half miles to six and a half
miles. Additional training included the use of gas masks and helicopter familiarization. These changes meant, with few exceptions, that
the WAC basic course was on a par with the basic course that men
underwent.[30]

In September 1975, Army Chief of Staff Frederick C. Weyand
directed that TRADOC develop a plan to consolidate basic training for
men and women. General DePuy, though seemingly reluctant, agreed
to accept the consolidated POI. TRADOC immediately began development of an experimental six-week common-core BIET program (POI
21–114). The training program, which was similar to the one that was
used for men, was completed in December 1975.[31]

As soon as the Army staff approved the pilot program, the TRADOC
commander began developing a test of the course, to be held at the
Army Training Center at Fort Jackson between 17 September and 11
November 1976.[32] Approximately 875 male and 825 female nonprior
service recruits assigned to the 6th and 7th Battalions, 2d BCT Brigade,
made up the control and test groups, respectively. Each training battalion had four companies, two female and two male. The cadre of each
platoon within the companies included two male drill sergeants and one
female drill sergeant. Efforts were made to balance the cadre regarding education, length of service, experience, and grade level. Of great
concern at the Department of the Army level was the possibility of public censure of the Army by the public and the press. The authors of
the WITA study, also ongoing at the time, wrote: "Because of potential

---

[30] Morden, pp. 362–63.

[31] Morden, p. 363; Program of instruction for basic training of male and female military personnel without prior service (six weeks), POI 21–114, test edition, Dec 1975.
Hereafter cited as POI 21–114, Dec 1975.

[32] In addition to the testing of the program of instruction, the test plan included
a requirement for ARI to conduct an attitude survey of both cadre and trainees. The
survey was conducted during BIET at Fort Jackson, S.C., in late 1976 but was not published until Sep 1978. Thus, the results of the survey were not widely available in the
early days of the mixed-gender experiment. *Basic Initial Entry Training Test Attitude
Survey* (Alexandria, Va.: ARI, Sep 1978).

public sensitivity to this issue and for the purposes of this test, male and female trainees must be assigned to separate training organizations."[33]

TRADOC exercised overall supervision of the BIET test. The test director was the director of training and education at the WAC Center and School at Fort McClellan; other support came from the Infantry School, the training staff at Fort Jackson and Fort McClellan, and the TRADOC Combined Arms Test Activity at Fort Leavenworth, Kans.

The objective of the BIET test was to compare the performance of male and female trainees against the standards of a common core of instruction to identify necessary changes.[34] As Maj. Gen. Robert C. Hixon, TRADOC chief of staff, wrote the Army staff in January 1977, shortly after conclusion of the BIET test: "The test represents the first instance where U.S. Army male and female personnel have been trained to the same skill level under the same conditions and their performance evaluated against the same standards at the entry level."[35]

Hixon wrote that the test had revealed little difference in the relative performance of male and female recruits except in the area of physical readiness. This was no surprise to those who had followed prior test results. Of special interest to TRADOC was the finding that training standards demanded of men had not been reduced to accommodate women; the POI had sought specifically to avoid compromising male training. Other observations were that female soldiers were not properly outfitted and required better boots, warmer field jackets, and undershirts, which were issued only to men. In addition, consolidated training would cost approximately $47 per woman for increased ammunition expenditures, because women would be trained on weapons not traditionally included.[36] In February, the DCSPER wrote to General DePuy to approve TRADOC's report and to request that the integrated course of instruction be implemented in February 1977. He also suggested that provisions for sex education and rape prevention instruction be included before final submission to the Department of the Army headquarters (HQDA) for approval.[37]

---

[33] WITA study, 1976, p. 8-7.

[34] Terms and acronyms for various facets of Army basic training vary from time to time and author to author. *Basic combat training* (*BCT*) refers to basic training for men (not including advanced individual training, or AIT) before the disestablishment of the WAC. WAC training was known simply as *BT* for *basic training*. The term *basic initial entry training*, *BIET*, or later *IET*, refers to BCT plus the skill and MOS phases of AIT.

[35] Ltr, TRADOC chief of staff to Department of the Army headquarters (HQDA) (DAPE), 5 Jan 1997, sub: *Basic Initial Entry Test* report.

[36] Ibid.

[37] Ltr, HQDA DCSPER to Cdr, TRADOC, 16 Feb 1977, sub: *Basic Initial Entry Training Test* report. The new consolidated training program for men and women also

Implementation of Integrated Training

On 16 February 1977, Army Chief of Staff Gen. Bernard W. Rogers approved initiation of consolidated basic training for men and women based on the BIET course of instruction and on the field test conducted at Fort Jackson. Plans called for the conversion of existing WAC training programs at Fort McClellan and Fort Jackson to the new program as soon as possible. The U.S. Army Infantry School at Fort Benning, Ga., the proponent for the existing BCT program, began development of an expanded version (eight weeks, including one fill week) of the test version used at Fort Jackson.

The revised training program for women at Fort McClellan gradually aligned female basic training with the male basic training program at Fort Jackson.[38] Integrated training began at Fort Jackson on 2 October 1977. Women were integrated down to the company level. Four basic training companies (about 200 trainees) had three male platoons and one female platoon (about 50 members each), while one company had two male and two female platoons. The draft course of instruction (POI 21–114) developed at the Infantry School began on 14 October 1977. The final version of POI 21–114 was introduced in March 1978. In the next phase of the integrated program, TRADOC officials informed Fort Leonard Wood, Mo., and Fort Dix, N.J., on 14 July 1978, that the two bases were to initiate training for women on 1 October 1978 and informed Fort McClellan that it would begin BCT for men on the same day.[39] Early in September, TRADOC approved the Fort Leonard Wood and Fort Dix plans for mixed-gender training. According to TRADOC's plans, the next phase of the program would include AIT, for women, beginning in the second half of fiscal year 1979 at all installations where such training was being conducted for men. That phase of implementation began with studies from Fort Knox, Ky., Fort Bliss, Tex., and Fort Sill, Okla. Men and women began integrated AIT at Fort Bliss on 21 November 1978 and at Fort Sill on 9 February 1979. TRADOC determined that the number of female AIT students did not support an integrated program at Fort Knox.[40]

---

included curriculum changes for the Army's proposed One-Station Unit Training program (OSUT), which would begin in 1977 and include women in some noncombat MOSs, especially Military Police and Signal positions. In the OSUT program, recruits learned some MOS skills while undergoing basic training at installations that also conducted branch AIT.

[38] TRADOC AHR.

[39] According to TRADOC records, training actually began at Fort Leonard Wood and Fort Dix on 13 Oct 1978 and at Fort McClellan on 27 Oct 1978.

[40] TRADOC AHR, FY 1977, pp. 77–80; FY 1978, pp. 58–60; FY 1979, p. 92; FY 1981, p. 209; FY 1982, pp. 220–22.

Thus, the Army joined the other services in training men and women together in basic combat training. The new program had, by Army standards, been rather hastily designed and implemented. However, given the social and political climate of the late 1970s, the service would seem to have had little choice.

The Experience

What was the experience of the women who were members of the first mixed-gender basic training units? The experience of the men in those units is of no less importance, but they left few personal records, perhaps because the changes for them were less remarkable and dramatic. On the other hand, the women were observed, subjected to numerous studies, reviewed, interviewed, and analyzed. This section focuses on no particular Army training center; it is, however, a composite of the women's experiences in the 1978–1981 period.[41]

The women who began basic training in integrated units in 1977 and 1978 had, for the most part, enlisted in the Army for the same classic reasons as male recruits. However, many were attracted by pay scales, benefits, and perceived promotional opportunities that—unlike in the civilian sector—were equal to those of men. The priorities for others were job training, educational opportunities, and funding for education. Some came from military families. Some had patriotic reasons for joining. Others were attracted by promises of security, adventure, and travel. Some young women simply did not know what they wanted to do or looked to the Army as an escape from an undesirable family situation or a small hometown. Whatever the reasons, it is clear that most looked at enlistment in the Army as an interlude in their lives. Few intended to make the Army a career.[42]

During basic training, recruits did not receive training in specific job skills, that is, in an MOS. Specialized training occurred in AIT in most cases. However, upon enlistment, each recruit chose or was assigned an MOS from those available. As of 1978, the Army's combat exclusion regulations barred women enlistees from choosing an MOS in Infantry, Armor, Cannon Field Artillery, Combat Engineer, or Low-Altitude Air Defense Artillery. Statistically, women could be assigned to 324 of the Army's 348 enlisted specialties. In practice, only 59 percent of the

---

[41] For this section, I owe much gratitude to Helen Rogan's *Mixed Company: Women in the Modern Army* (New York: Putnam, 1981). Rogan's study is based, in part, on her participation in a six-week training cycle at Fort McClellan in the spring of 1979.

[42] Dorothy and Carl J. Schneider, *Sound Off: American Military Women Speak Out* (New York: E.P. Dutton, 1988), p. 5; Rogan, pp. 49, 60–61.

actual spaces available were open to both men and women because of the heavy concentration of personnel in combat arms positions. Another trend regarding the choice of MOS that especially concerned the Army leadership was the tendency of a majority of women recruits to choose traditional specialties (clerical and administrative) to the neglect of mechanical, technical, or law enforcement fields. If the service was to make the best use of increasingly female strength, it needed many more women to contract with recruiters for nontraditional fields.[43]

Upon arrival at the training center, the new recruits, male and female, were usually housed in the same barracks but not on the same floors. In some cases, more restroom and laundry facilities had been added to the women's floors, along with shower curtains and partitions. With the increased emphasis on physical training, field training, and weapons familiarization, the women's training uniform changed from a three-piece field uniform to olive-green fatigues, M-1 helmet liners, and men's combat boots, which were soon replaced with boots designed especially for women—specifically for nurses, who spent much time standing on concrete floors. For a trainee in the integrated Army, women's boots were poorly fitting and contributed to blisters, shin splints, stress fractures, and other injuries, leading some to claim that the boots prevented women from functioning at full capacity. In May 1979, the Army abandoned them for a return to male boots that had never really proved satisfactory either.

A white t-shirt, accompanied by an olive-green shirt or "jacket" in colder weather, and a green baseball cap completed the training duty uniform. The women's fatigues, at least at Fort McClellan, were not of a wash-and-wear fabric the men's were, but were made of a cotton fabric that required starching and pressing. Meanwhile, tests were ongoing to measure and issue smaller helmets and protective vests for women soldiers.[44]

The development of a POI for mixed-gender training at the Army's basic training centers has been described above. Likewise, the changes in the POI at the previous WAC training center at Fort McClellan to include additional weapons training and increased physical conditioning has been noted previously. The new POI for the formerly all-male training centers remained basically the same, with one notable exception: A number of hours of training were set aside for sex education

---

[43] Department of the Army Historical Summary, FY 1979 (Washington, D.C.: Center of Military History), p. 66; FY 1980, p. 95. Hereafter cited as Army Historical Summary. The Army grouped MOSs under three headings: traditional, less traditional, and nontraditional.

[44] Morden, p. 363; Rogan, pp. 36–37.

and rape prevention instruction, as directed by the Department of the Army. That action resulted in a decrease of hours for the largest block of instruction, M-16 rifle marksmanship. With regard to tasks and instruction, all training was coeducational and the new POI made no reference to gender.[45]

Although the choice or assignment of MOS did not directly affect the common-core instruction of basic training, that choice became embroiled in issues regarding the second largest block of instruction: physical training. The increasing concern about the physical ability of women to perform certain military tasks has been noted. With the certainty that women would receive basic training in integrated units, TRADOC began studying the physical requirements associated with each MOS to develop gender-free physical standards that would determine who could perform effectively in each specialty, regardless of gender. In November 1978, Armed Forces Examining and Entrance Stations began using an "X-factor" test, which related an individual's weight-lifting ability to the physical requirements of a particular skill. The test, however, was ultimately used only to advise enlistees—both male and female—of the chances of success in their chosen MOS. No one was denied entry into a specialty on the basis of the test alone. By January 1980, Chief of Staff Gen. Edward C. Meyer and TRADOC commander Gen. Donn A. Starry had agreed that the MOS-related system was too complex and tended to lower standards. TRADOC then began to develop a standard physical training test for all soldiers, but with standards adjusted for the physiological differences between men and women. That decision was based on the belief that the existing training system challenged women but not men and, therefore, had to be toughened. The new physical training program went into effect in October 1980, featuring only push-ups, sit-ups, and a two-mile run.[46]

Another important training module, M-16 A1 basic rifle marksmanship, also brought into question the physical abilities of many women encountering a formerly all-male POI. Prior to 1977, the M-16 A1 training course for the WAC had been limited and voluntary. In this year, the course became mandatory, and successful completion was required for graduation from BCT. Trainees were taught safety, cleaning and assembly of weapons, and automatic and semiautomatic fire and were prepared to qualify on a 25-meter range using silhouette targets.[47] These

---

[45] TRADOC AHR, FY 1977, p. 79.

[46] TRADOC AHR, FY 1979, p. 66; FY 1980, p. 45; FY 1981, p. 48.

[47] It was Army policy at this time that no pictures were allowed of women training with rifles out of a concern for the possible effect on public sensibilities and questioning of the combat exclusion regulation. Rogan, p. 51.

skills proved no more difficult for women than for men. The M-16 A1 weighed about six and a half pounds. The problem for many women was seeing over the sandbags that buttressed the foxholes. Standing on boxes and crates proved only a partial solution.[48]

During an increasingly busy training schedule, recruits spent a number of hours in mixed-gender classrooms. There, they received instruction on the proper way to wear the uniform, the responsibility of being a soldier, and the role of the Army (military history). One observer of the instruction at Fort McClellan noted that orientation films showed only male soldiers and old WAC tapes were used as training aids.[49] Trainees were also introduced to military justice, the code of conduct, and the rules of the Geneva Convention and took part in a seminar on race relations. All or some of these blocks of instruction were often conducted by a judge advocate general, who lectured on the treatment of prisoners of war and the required protocol and procedures if one was captured. Classroom sessions also included instruction in the avoidance of venereal disease, the use of contraception, first-aid, drug and alcohol abuse, and the prevention of heatstroke and frostbite.[50]

Field training included a wide variety of events, such as drills, ceremonies, and instruction in courtesy and customs. Drill sergeants paid special attention to ensure that recruits recognized musical selections important to the Army: "The Star-Spangled Banner," "The Army Goes Rolling Along," "Hail to the Chief," "To the Colors," "Reveille," and "Retreat." Training also included chemical and biological defense instruction using "tear gas." Early on in mixed-gender training, only the protective mask was used; later, procedures for the entire protective suit were added. In addition, trainees were introduced to basic weapons that they might be called upon to fire in an emergency situation, such as the M-60 machine gun, the M-203 grenade launcher, the light antitank weapon, and the Claymore anti-personnel mine. All recruits also were required to negotiate a hand-grenade assault qualification course and demonstrate the proper way to throw a hand-held grenade.[51]

Perhaps the most dreaded events were the road marches and the time spent in temporary living quarters, known as bivouacs; individual tactical training; and the legendary obstacle course. Bivouac for a minimum of four consecutive nights and road marches of seven, ten, twelve, and fifteen miles, carrying the M-16 A1 rifle and field equipment, were required. The entire load weighed almost thirty-five pounds. Individual

---

[48] Ibid., p. 54.
[49] Ibid., p. 93.
[50] POI 21–114, Dec 1975.
[51] Ibid., passim, POI 21-114, Dec 1975.

tactical training in offensive and defensive techniques, often conducted during bivouac, was originally included in the basic training POI for mixed companies. The training included the use of cover and conceal-ment (face paint and camouflage foliage), the low crawl, high crawl, and the digging of foxholes. Bayonet training and the use of barbed wire had recently been eliminated. Tactical training in offensive tech-niques for women was soon eliminated.[52]

Without a doubt, the most memorable basic training program was the obstacle course, required for graduation from BCT. A common program was followed for one to two weeks, during which trainees were divided into three groups or tracks: unconditioned, medium-conditioned, and well-conditioned. The goal was to move personnel to the next highest level. That arrangement made improvement the criterion for success rather than a prescribed performance standard. Tasks included climbing over and under logs, climbing ropes, crossing a balance beam, scaling walls and fences, jumping across a pit, climbing hand over hand, and climbing the high cargo net. Injuries were common even before gender integration. A conditioning course—practice on each obstacle in t-shirts and run against time—preceded the ultimate "confidence course." The final test was not run against time and re-tests were allowed.[53]

That, then, was the structure of the basic training program when the first male and female integrated units began arriving at Army training centers in 1977 and 1978. Simply put, the reactions of the young train-ees and the training cadre to the "new integrated Army" were varied and individual. Some women believed that a pervasive attitude prevailed among most male recruits that women were physically inadequate and did not belong in the Army. Others sensed that women had become associated with all that was wrong with the post-Vietnam Army. Some female recruits believed they were being forced to earn, as a privi-lege, a place in the military that men either claimed or rejected as a birthright. Some, perhaps a majority, of male trainees believed women received preferential treatment and should be required to contribute their share.[54]

Not all attitudes were negative. In some ways, BCT seemed to have more to do with a unique experience than with men, women, or gender integration. On balance, most trainees accepted the

---

[52] Ibid., passim, POI 21-114, Dec 1975; Schneider, pp. 30–31.

[53] POI 21-114, Dec 1975, pp. 25–26, 36.

[54] These composite "attitudes" are derived from Rogan, *Mixed Company*, which is based on her experience during an entire training cycle at Fort McClellan, and Schneider, who conducted more than 300 interviews. The latter are stored at Schlesinger Library, Radcliffe College, Cambridge, Mass.

status quo. A former trainee at Fort McClellan in 1979 later described her experience: "Probably half of the women who went through that program wish that it had stayed integrated and half of them think it's better now. I wish it was the way I went through [integrated]. Unless the men actually see you doing those things, they don't believe it."[55]

A female lieutenant of the Fort McClellan BCT cadre ran her company on the assumption that the most important differences were between people: "Everyone keeps telling me there are differences, but I don't see them. They say the men can't march and the women can; the men can't houseclean and the women can; the women can't fire well and the men can....The variations always had more to do with the drills than the trainees."[56]

A male trainee gave his opinion: "A lot of us think the women slow us down—but still, it helps a lot just having them around. You can talk to them in a way you can't talk to men."[57]

Another observer summed it up this way: "Army women are required to become effective soldiers, in just the same way that the men are, and the process leads us to question all our traditional ideas about masculinity and femininity."[58]

The Problems

Although TRADOC and the Army publicly declared the post-WAC, mixed-gender basic training program a success, from the beginning, there were problems—minor and major, anticipated and unanticipated. Some of the problems with regard to M-16 A1 rifle marksmanship, physical training, and uniforms have been noted. In addition, from the start, there was never a sufficient number of drill sergeants—especially female drill sergeants—a problem that affected training, as well as the supervision of housing arrangements. The leadership and supervision of women by men trained to lead only men caused concern among trainers and trainees alike.

There were also difficulties regarding recruitment, enlistment criteria, and attrition. In the early years of the AVF, female recruitment goals were achieved with relative ease. However, by 1978 and 1979, the combined effect of increasing annual female recruiting objectives, the denial of enlistment to women without high school diplomas, the requirement that women enlist in nontraditional skills, as well as a tougher

---

[55] Schneider, p. 31.
[56] Rogan, p. 42.
[57] Ibid., p. 104.
[58] Ibid., p. 29.

recruiting environment, created unexpected difficulties. By 1986, the Army's goal for the number of women on active duty, which was set by the Department of Defense at the urging of the Carter administration in 1977, was 87,500, up from 62,000 in 1980. In addition, male recruits were not required to have high school diplomas. Although President Carter's secretary of the Army, Clifford L. Alexander Jr., lowered the maximum score required on the Armed Forces Qualification Test from 59 to 50 and again from 50 to 31, recruiting of women became increasingly difficult. In FY 1978, the active Army reached 99.5 percent of the female recruiting objective; the following year, it achieved only 91.5 percent of the goal.[59]

Effective 1 October 1979, enlistment eligibility criteria for men and women were standardized in an attempt to remedy the recruitment shortfall. Thus, female non-high school graduates could enlist with a minimum score of 31 on the entrance test. The new rules caused a substantial surge in female enlistments, but the year 1980 ended with accessions still about 1,000 short of the goal. The new criteria also exacerbated what was already a major problem among enlisted women: a rising attrition rate in the first enlistment term. Statistics indicated that the three-year attrition rate for first-term female soldiers was up to 20 percent higher than for their male counterparts and was increasing. The numbers in FY 1980, the first year of the new enlistment equality program, indicated that 46 percent of non-high school graduate males failed to complete three years of service, while the number for females was nearly 62 percent. In FY 1981, first-term attrition of recruits without a high school diploma was 50 percent for men and 68 percent for women. In addition, some among the Army leadership believed that the caliber of recruits was lower among non-high school graduates, a factor that contributed to other problems regarding mixed-gender training at the basic level and further complicated the debate about the utilization of female soldiers. From the vantage point of the female recruits, the increasing problems led some to abandon their hopes and expectations for an Army career.[60]

As the Army and TRADOC sought to make whatever changes seemed indicated to ensure the success of integrated training of men and women in BCT, other events took center stage. In the summer of 1978, two serious abuse incidents occurred during basic training. The first incident, at Fort Jackson, resulted in the death of two male trainees from heatstroke on the first day of training. General court-martial

---

[59] Army Historical Summary, FY 1980, pp. 95–96.
[60] Army Historical Summary, FY 1981, p. 81; FY 1982, pp. 49–50.

proceedings found two male drill sergeants guilty of involuntary manslaughter, negligent homicide, and dereliction of duty. The other incident, at Fort Dix, involved five drill sergeants in the systematic abuse of twenty-two trainees. Three of the five trainers were convicted of trainee maltreatment. Subsequent publicity led to hearings by the Investigations Subcommittee of the House of Representatives Armed Services Committee. Testimony on the incidents came from Secretary of the Army Alexander, Chief of Staff Bernard W. Rogers, TRADOC commander Gen. Donn A. Starry, and the commanders of the basic training centers.

This tragedy focused the TRADOC leadership's attention once again on the continuing problems with basic training that had been a matter of concern with the command since its formation. In July 1978, TRADOC commander Starry, concerned that these trainee abuse incidents indicated the existence of widespread and deep-seated leadership problems in the training base, established a special task force to conduct a detailed study of the matter.[61] The Task Force on Initial Entry Training Leadership published its report, known as the Raupp report, named after Col. Edward R. Raupp, who served as chairman in December 1978. The committee found that basic training was tough, demanding, and stressful and that trainers were producing soldiers who were highly motivated. Major incidents of trainee abuse were found to be the exception rather than the rule. However, the study revealed significant differences in basic training among TRADOC's training centers, a situation encouraged by the lack of a centralized TRADOC policy.

As a result of the Raupp report, General Starry established a committee composed of commanders from the nine training centers to examine the problems and make recommendations to standardize procedures in IET. The Committee of Nine, as it was known, was chaired by Maj. Gen. Robert C. Hixon, TRADOC chief of staff.[62] The Committee of Nine submitted its report in July 1979 to Starry, who issued a statement of training practices for the centers on the basis of the report's recommendations. His statement emphasized that positive leadership in basic training discouraged an atmosphere conducive to abuse of individual trainees. Whatever the outcomes of the studies, the incidents at

---

[61] The Initial Entry Training Task Force of 1978 was concerned not only with BCT but with AIT and the new OSUT program.

[62] In 1974, TRADOC's first commander, Gen William E. DePuy, set a precedent for involvement of the training center commanders in finding solutions to problems when he established the Committee of Six, composed of the commanders of the Army training centers. The committee was chaired by TRADOC deputy commander, Lt Gen Orwin T. Talbott. TRADOC AHR, FY 1982, p. 217.

the Army's training centers generated a substantial number of critical media reports and invited close public scrutiny of the new mixed-gender training for Army recruits.[63]

As the Army and TRADOC sought solutions to these alleged incidents of trainee abuse, the service encountered its first major experience with charges of sexual harassment since the advent of the AVF. The aforementioned *Evaluation of Women in the Army* study of 1978 had made it clear that sexual harassment was not to be tolerated, although there were no plans to address the subject in detail. The Army would continue to support a previous policy statement that directed commanders to educate soldiers on the subject and enforce existing policies regarding harassment. Generally, with full integration, harassment seemed to become much more common or, at least, more visible as women became bolder about reporting possible occurrences.

Then, beginning in December 1979, the *Baltimore Sun* published a series of articles about alleged harassment at Fort Meade, Md. The articles painted a graphic picture of behavior toward women—especially those of the lower ranks—that was seen as sexual harassment. Soon, investigations were reported from Fort Benning, Fort Dix, Fort Bragg, N.C., and the Presidio in San Francisco, Calif. On 4 January 1980, Secretary of the Army Alexander and Army Chief of Staff Edward C. Meyer sent a joint message to the field reaffirming the Army's commitment to "a policy that upholds the human dignity of all military and civilian personnel." Meyer also directed the inspector general to investigate allegations of sexual harassment and mistreatment of women at a number of Army installations. In addition, the House Armed Services Committee held hearings on the matter in February 1980. Testimony of soldiers stationed at Fort Meade, ten female enlistees selected at random, and the top-ranking women in each of the services seemed to indicate, predictably, that sexual harassment was most often practiced against young enlisted women by senior enlisted men.

Meanwhile, it had become increasingly clear to the Army leadership that sexual harassment contributed significantly to the high rate of female first-term attrition. In May 1981, Secretary Alexander signed a memorandum for all personnel defining *sexual harassment* as improper influence in exchange for sexual favors or offensive comments, gestures, or physical contact of a sexual nature in a work or duty environment. He reminded recipients that sexual harassment was unacceptable

---

[63] Army Historical Summary, FY 1979, p. 25. For a detailed description of the work of the Raupp committee and the Committee of Nine, see TRADOC AHR, FY 1979, pp. 105–118.

and incompatible with professional behavior. In August 1981, in the early months of the Reagan administration, Secretary of Defense Caspar Weinberger declared sexual harassment to be unacceptable conduct that would not be tolerated. Whatever the case, the heightened public awareness of the harassment of female recruits created a hostile environment as the Army leadership struggled with other issues confronting mixed-gender training.[64]

As the problems at the Army training centers increased, morale among men and women trainees suffered. In 1980, the situation worsened, and debate on the proper role of women in the military services became more heated when President Carter included both women and men in his plan for peacetime draft registration. Suddenly, the Army's morale problems were highly publicized. In defeating the proposal, the Senate Armed Service Committee's report called the proposal a smokescreen to divert attention from the serious manpower shortage in the AVF. Carter's effort to include women in draft registration failed, but Army Chief of Staff Meyer initiated efforts to improve the morale of the troops, especially new trainees. Congress approved additional funds for pay raises, enlistment bonuses, and a variety of increased benefits. Recruits in basic training were given insignia denoting the branch that they had chosen. Tests began on changes in women's uniforms. By early 1981, unisex fatigues, as well as a special version for women, were under development. New helmets made of a plastic-like material called Kevlar, were being tested as possible replacements for the steel helmet. Design changes were also made on the bulletproof vests and field pack harnesses that were issued to women.[65]

Despite these changes, the Army's senior leadership was acutely aware that serious problems continued in the newly integrated basic training system. Some leaders privately, and a few publicly, called for more studies to identify what other changes could be made and to determine what new directions were needed to ensure the Army's combat readiness. During its 1981 fall conference, the Defense Advisory Committee on Women in the Services (DACOWITS) expressed deepened concern for the morale of women in the military and called on the military services and the Department of Defense to provide "succinct comments on current and future remedial actions."[66] Others worried that critics of mixed-gender training at the basic level were accusing the Army of emphasizing social experimentation rather than national

---

[64] Stiehm, pp. 144–45, 205–06; Rogan, p. 242; Army Historical Summary, FY 1980, quotation, p. 96; FY 1981, p. 98.

[65] Rogan, pp. 229–31.

[66] DACOWITS History of Recommendations, fall conference 1981.

security. As one student of the subject noted: "Army leaders often found themselves in no-win situations. If a new policy broadened opportunities for women, it would be attacked by conservatives and military traditionalists with charges that the Army was going too far—catering to feminists at the expense of readiness. If a policy reduced opportunities, it was challenged by [the Office of the Secretary of Defense], DACOWITS, and the press as being regressive and antiwoman. The Army found no middle ground."[67]

But the military leadership was also aware that studies were unlikely to effect many changes so long as the Carter administration, with its unabashed support for women's rights and opportunities, occupied the White House and maintained a strategically placed staff at the Pentagon.

The Election of 1980 and Other Changes

Then, in the presidential election of the fall of 1980, challenger Ronald Reagan scored a landslide victory over the incumbent, President Jimmy Carter. The Reagan campaign had been based on socially conservative principles and an emphasis on the need to build U.S. defenses. The military leadership, especially in the Army, had every reason to believe that their concerns about training, the proper utilization of women, and combat effectiveness would receive a more sympathetic hearing than they had during the Carter administration. Further, the changes were not confined to the executive branch. Legislatively, it had become clear that the ERA, minus any mention of women and combat exclusion, would not be successful in enough state legislatures to become a part of the Constitution. Only a few years earlier, all the military services had been certain the amendment would pass, and a number of their actions and initiatives had been based on that assumption.

Finally, there was a shift in the legal winds. Traditionally the federal courts had been relatively strong supporters of expanded opportunities for women, in and out of the military services. On 26 June 1981, that trend was slowed when the Supreme Court overturned a lower court's decision and ruled in *Rostker* v. *Goldberg* that Congress had acted constitutionally in excluding women from a new draft registration program. As previously noted, that program had been enacted in the summer of 1980 as part of the U.S. response to the Soviet invasion of Afghanistan. The Carter administration had unsuccessfully sought to require the registration of both men and women. Subsequently, opponents of the draft

---

[67] Holm, pp. 198–99.

had filed a due process challenge against the constitutionality of an all-male registration. The three to six majority opinion rested on the argument that the intent of Congress, in authorizing the draft, was to provide Americans suitable for fighting as combat troops. Women, because of congressional and military restraints on their use in combat, were not suitable for purposes of the draft or registration for a draft.[68]

During this transition time, the Army and TRADOC headquarters made few public statements regarding the problems in the integrated training base. In fact, they had, on a number of occasions, publicly called the new program a success and encouraged field commanders to provide "knowledgeable, understanding, affirmative and even-handed leadership to all our soldiers."[69] This position was not held by field commanders, who were quite vocal in detailing their complaints and in questioning the effects of women soldiers on the readiness of their units. They were given a number of opportunities to file lengthy and detailed reports and surveys.[70] For example, the EWITA study of 1978 had been based on extensive surveys of field commanders. The Raupp committee on trainee abuse and TRADOC's Committee of Nine of late 1978 and 1979 had generated numerous reports. Similarly, the Gang of Four established by TRADOC commander General Starry in June 1981 had a charter to serve as a forum for commanders with similar interests and problems concerning IET.[71] Despite the availability of so much information, most of which revealed serious concerns, it was not until the political climate changed that the leadership believed it could—and should—reconsider mixed-gender training in BCT.

The Return of Separate Training

Even before Ronald Reagan was sworn into office in January 1981, the military services, jointly, sent a confidential position paper to his transition team asking him not to implement the "arbitrarily" large increases in planned accessions inherited from the Carter administration. The memorandum cited the need for total reassessment until the

---

[68] Michael Rustad, *Women in Khaki: The American Enlisted Woman* (New York: Praeger Special Studies, 1982), p. 92; Army Historical Summary, FY 1981, p. 96. For a more detailed treatment of this case, see Holm, pp. 373–78.

[69] Msg, Chief of Staff Bernard W. Rogers to all U.S. Army reps and activities, 032227Z Mar 1978, sub: Women in the Army.

[70] TRADOC Cdr. Starry to the Department of the Army, 041819Z Aug 1978, sub: Women in the Army.

[71] For a detailed account of the Gang of Four, see TRADOC AHR, FY 1982, pp. 215–20.

services reestablished their requirements and to avoid threats to established standards. The services requested that they—not civilians—be allowed to determine end strength and the proportion of women needed.[72]

In late February 1981, with a change in presidential administrations, the Army indicated that it would abandon the personnel goals of the Carter administration and maintain the current end strength for enlisted women soldiers at sixty-five thousand to allow time for a thorough study of the problems discussed above. Acting Assistant Secretary of the Army for Manpower and Reserve Affairs William Clark went before the Senate Armed Services Committee to announce the new policy and the service's plans for an extensive study on the impact of female soldiers on the Army's combat readiness. The service referred to its requested policy as a "pause" in recruitment and accessions, a phrase that later evolved into the term *womanpause*.[73] At that time, Army leaders explained that reports from field commanders questioning the effects of women soldiers on the readiness of their units would be used to evaluate the current situation.[74] It was not a question of women doing their jobs or of exposure to combat, Clark explained, but rather "the combat effectiveness of the organizations as you have large numbers of women in them." Some students of the Army's action at that time in freezing recruitment goals and accessions believed that the agenda was to force a return to the draft by undercutting personnel for the AVF.[75]

In March 1981, as more questions were asked both in and out of the military about the relative merits of "social experimentation" and combat effectiveness, the Department of Defense—under the direction of Deputy Secretary of Defense Frank C. Carlucci—began a joint background review of the impact of present and projected numbers of women in the services.[76] In May, under orders from Chief of Staff Meyer, the Army followed suit with the establishment of a WITA Policy Review Group, as requested by Congress. The group was scheduled to present its initial findings to the chief of staff by 31 December 1981. (The activ-

---

[72] Stiehm, pp. 54–55. The total of women in the military numbered about 150,000 in Jan 1981. The Carter administration goals were for 223,000 by 1986.

[73] Jeffrey M. Tuten, "The Argument Against Female Combatants" in *Female Soldiers: Combatants or Noncombatants: Historical and Contemporary Perspectives*, Nancy Loring Goldman, ed. (Westport, Conn.: Greenwood Press, 1982), p. 262. It is not clear whether the Army informed DoD of the expected appearance in Congress before the hearings. It appears DACOWITS was not informed.

[74] Army Historical Summary, FY 1981, p. 96.

[75] Holm, pp. 388, 390.

[76] This review was completed in Oct 1981; Mitchell, p. 102; Holm, pp. 393–97; Stiehm in Loring, p. 190.

ities of this group are discussed below.) By May, the freeze on recruitment by the Air Force and the Army was official. Neither the Navy nor the Marine Corps indicated a change in plans. Assistant Secretary of Defense for Manpower, Reserve Affairs, and Logistics Lawrence J. Korb explained: ". . . [N]ow, we have some experience. I think it's an appropriate time, at the beginning of an administration when you are having a force expansion, changing doctrine, to take a look and say, OK, let's stop and see if these models should be changed....maybe we were a little bit too eager, and [doing] a little too much, maybe wishful thinking."[77]

Meanwhile, the Army staff and field agencies began to focus on the status of women in the Army. The ARI initiated a study to determine the causes of the differences in male versus female attrition rates in the Army's approximately 350 enlisted MOSs. The Army Audit Agency conducted a similar survey. Both studies were scheduled to be completed in early 1982.[78]

In June 1981, the Army abandoned the enlistment eligibility criteria that had been standardized in 1979 for men and women and enlisted only a few women who were high school graduates. In fact, all Army recruitment of women was virtually halted for the remainder of the fiscal year. At the same time, in addition to awaiting the results of several major studies at the Department of Defense level, TRADOC had begun making changes in the basic training curriculum, hoping to resolve some motivational problems by more and better training. In January 1981, the command implemented an eight-week program of instruction at Fort Knox and Fort Leonard Wood; in October 1981, the new program began at Fort Dix, Fort Jackson, Fort McClellan, Fort Sill, Fort Benning, Fort Bliss, and Fort Gordon, Ga. The revised POI, developed at Fort Benning, eliminated twenty-four hours of existing training in favor of reinforcement training. The hours eliminated included eleven from individual tactical training, three from marches and bivouac, five from drill and ceremonies, two from military communications, one from opposing forces orientation, and two from U.S. weapons.[79] Although the changes in the basic training program affected both male and female trainees, it is clear that the longer training cycle and increased skill reinforcement were focused, at least in part, on a possible solution to some

---

[77] As quoted by Tuten, p. 262, from the *Washington Post*, 13 May 1981.

[78] Army Historical Summary, FY 1981. The author was unable to locate a copy of the ARI report or the Audit Report HQ-82–212, *Enlisted Women in the Army*, Apr 1982.

[79] TRADOC AHR, FY 1982, pp. 220–21.

of the problems that many senior leaders believed were the results of mixed-gender basic training.

Although 1982 would prove to be an uncertain year for women in the military, especially for the Army's new recruits, early in the year, the Reagan administration warned that it would not allow a moratorium on women's roles in the armed forces. In mid-January, Secretary of Defense Weinberger directed the military services to "aggressively break down those remaining barriers that prevent us from making the fullest use of the capabilities of women in providing for our national defense." In other words, the military services should not misread the intentions of the Reagan administration nor should they expect any consideration of a return to the draft. Deputy Secretary Carlucci directed all the service secretaries to report on actions to remove institutional barriers to women's military service that might mean discrimination in recruiting or career opportunities. In March 1982, Assistant Secretary Korb officially ended the Army's "pause" in recruitment and enlistments by announcing that the numbers of women in the armed forces would continue to rise in the years ahead.[80]

Then, two months later, in May 1982, Army Chief of Staff Meyer suddenly and quietly announced that the service would discontinue coeducational basic training by 30 August 1982, but the date was later changed to October 1982 to accommodate those already in training. Subsequently, TRADOC commander Gen. Glenn K. Otis, who replaced Gen. Donn A. Starry on 1 Aug. 1981, directed that women no longer be assigned to male platoons. Although the basic training POI would remain the same for both sexes, male and female recruits would be segregated at the company level and below. In short, instead of training in all-female platoons within a mixed company, women would be trained in all-female companies. In another change, women would receive basic training only at Fort Jackson, Fort Dix, and Fort McClellan. Women scheduled for basic training at Fort Leonard Wood would be phased into these three locations. Females programmed to enter OSUT at Fort Sill, Fort Bliss, and Fort Leonard Wood would enter the courses as mid-cycle fills after completing basic training at one of the three remaining basic training centers. Despite charges that the resulting program would be "gender pure," gender would not affect training cadre assignments; that is, male and female drill instructors would continue to train both men and women.[81]

---

[80] Army Annual Historical Summary, FY 1982, quotation, p. 58; Holm, pp. 395–97.

[81] TRADOC AHR, FY 1982, pp. 221–22; Army Historical Summary, FY 1982, pp. 58–59.

Although neither the Army, nor the TRADOC leadership, ever officially announced the reason for the abandonment of the mixed-gender basic training program in place since 1978, the informal explanation was that men were not being physically challenged enough in integrated training and were, therefore, not reaching their full potential. Single-sex companies, it was believed, would toughen the men and enhance the "soldierization" process. Spokesmen were quick to point out, however, that women had not failed to meet the Army's standards. The director of IET, in an interview with the *Army Times*, expressed the opinion that a training environment that included women had potential for an adverse psychological impact on new male soldiers. All-male companies, he went on to say, improved esprit de corps, encouraged competition, and pushed trainees beyond minimum standards.[82] The Army's decision regarding basic training was a surprise to almost everyone, although "back channel" memoranda had been circulating for several months. It is clear from the records that DACOWITS had no prior notice, nor did the Department of Defense announce the action.[83]

Opinions among the women most affected—those in basic training, those who would be, and female soldiers in general—are difficult to assess. Some women welcomed at least a partial return to what they perceived as the superior values and standards of WAC training. Others welcomed an end to the necessity to constantly compete with men. Enlistees who had been given no choice as to a traditional or nontraditional MOS welcomed the easing of pressure to accept nontraditional jobs. But an increasing number of Army women believed they were the subject of too many studies aimed at justifying decisions already made to restrict opportunities for women or to roll back gains made in the 1970s. Maj. Gen. Mary Clarke, the last WAC director and a seasoned veteran of "gender issues," called the Army's decision . . . [T]he first step toward returning women to their old stereotypical roles in the military....The next thing is that they will start closing military occupational specialties to women....The women are being let down.[84]

---

[82] Army Historical Summary, FY 1982, p. 58; *Army Times*, 17 May 1982. One observer reported, "According to a senior military official at the U.S. Army Training and Doctrine Command who was involved in the decision-making process on this issue, the primary reason for ending sex-integrated training was the perceived problem of fraternization between female and male soldiers." Melissa S. Herbert, *Camouflage Isn't Only for Combat: Gender, Sexuality, and Women in the Military* (New York and London: New York University Press, 1998), p. 16.

[83] The published recommendations from the spring and fall meetings of DACOWITS for 1981 and 1982 show no awareness of what doubtless would have been a major issue.

[84] *Pittsburgh Press*, 17 Jul 1982, as quoted by Stiehm, p. 62.

Almost all observers, opponents and proponents of coeducational training alike, remarked on the increasingly low morale among female soldiers.

The 1981–1982 Women in the Army Policy Review

With basic combat training once again "gender pure," there remained one piece of unfinished business regarding women in the Army: the report of the aforementioned WITA Policy Review Group. As noted above, the study group had been officially appointed in May 1981 with the mission to examine the effect of the enlisted female soldiers on the Army's combat readiness and effectiveness.[85] The panel was directed to report its initial findings to the Army chief of staff by mid-September, with the final report to be submitted by 31 December 1981. The DCSPER office, then headed by Lt. Gen. Robert G. Yerks and, after August, by Lt. Gen. Maxwell R. Thurman, was responsible for the work of the group.[86] The group's director, initially, was Maj. Gen. Robert L. Wetzel. He was assisted by five officers from DCSPER, an enlisted woman, a female civilian, and an eighteen-member special duty committee.[87]

Specifically, the review attempted to address, rather belatedly, problems identified by a General Accounting Office (GAO) report of May 1976 on job opportunities for women in the military. The GAO report notified Congress of increasing concerns that women were being assigned to specialties "without regard to their ability to satisfy the specialties' strength, stamina, and operational requirements." As always, the questions of the utilization of women and the differences in physical strength between men and women were tightly entangled with the prohibition against combat service and the debate over the definition of *combat*.[88] From the beginning, the study attracted an unusual amount of media attention and was looked on by many female soldiers and women's advocacy groups with suspicion. There were fears that the

---

[85] Plans for such a study group had clearly been in the works for several months before its formal announcement.

[86] Memo for the director, Women in the Army Policy Review Group, HQDA DAPE-ZA, 4 May 1981, sub: Tasking. Gen Thurman served as DCSPER from Aug 1981 to Jun 1983. From Jun 1983 to Jun 1987, he was vice chief of staff of the Army. From 29 Jun 1987 until 1 Aug 1989, Thurman was commander of TRADOC. Subsequently, he served as commander in chief of U.S. Southern Command during Operation JUST CAUSE in Panama.

[87] M.C. Devilbiss, "Women in the Army Policy Review: A Military Sociologist's Analysis," *Minerva*, fall 1983, p. 90.

[88] GAO Report FPCD-76-26, "Job Opportunities for Women in the Military: Progress and Problems," May 1976, p. 13.

services, especially the Army, were insincere in their support for gender integration.

Concurrent with plans for the WITA Policy Review, in April 1981, the study group directed ARI to study the length and frequency of absences of first-term male (393) and female (345) soldiers from their regular jobs (these soldiers had already entered their career management fields). The assumption was that women were absent from duty much more often than men and that losing time deprived young soldiers of the training and experience they needed to perform their jobs satisfactorily, with a resulting negative impact on future unit readiness. There was also concern that women's absences created resentment among male soldiers. Researchers obtained data on lost time from a five-day log kept by the soldiers' supervisors. Reasons for lost time were categorized as medical and health concerns, home and family care, and discipline-related actions. The preliminary data, available in the fall of 1981 but not released to the public, showed that two-thirds of the total number were away from their jobs for some time during the five-day period. Women were absent for health reasons more than twice as often as men (24 percent to 11 percent). Absences for home-related reasons and disciplinary actions were slightly greater for women than for men, but the differences were not considered significant.[89] For all lost-time categories combined, the amount of time away was about the same.[90]

Meanwhile, the DCSPER directed TRADOC and its commander, General Glenn K. Otis, who replaced Starry in August 1981, to develop a "methodology for assessing what women should do and where they should serve in the Army." The TRADOC model, which Army Chief of Staff Meyer approved in July 1981, became the responsibility of the Soldier Support Center (the reorganized and renamed Army Personnel Center) at Fort Benjamin Harrison, Ind. The TRADOC methodology, as applied to the existing force, was designed to determine physical demands for all enlisted MOSs and to determine the probability of exposure to direct combat for soldiers in positions in both TOE and TDA units. The necessary data would be collected from all TRADOC service schools. Completion date for the project was 1 February 1982.[91]

---

[89] The data on disciplinary absences is inconsistent with other studies.

[90] This result was, in part, due to the lower number of women in the study. Joel M. Savell, Carlos K. Rigby, and Andrew A. Zbikowski, "An Investigation of Lost Time and Utilization in a Sample of First-Term Male and Female Soldiers," ARI Technical Report 607, Oct 1982.

[91] Msg, Maj Gen Blount, TRADOC chief of staff, to distr, 181515 Aug 1981, sub: Implementation of TRADOC Methodology for Assessing What Women Should Do and Where They Should Serve in the Army.

By August 1981, those responsible for the review had extended the completion date into 1982. As time went by, suspicion of the study group's activities increased. In February, General Thurman announced that the WITA Policy Review Group's report would not only be delayed again, but its parameters had been severely restricted. Of nineteen issues identified for the original agenda, only issues specific to women would be addressed—pregnancy, the relationship between combat service and Army MOSs, and physical strength requirements of each specialty. The remaining issues, which included fraternization, assignment of military couples, single parents, sexual harassment, attrition and retention, privacy, and field hygiene, would be resolved within the Army's staff agencies. It is not clear whether the ARI data or the study on lost time was used by the study group, but neither was part of the final report and neither was released to the public until October 1982.[92]

Concern about the direction the Army's review study might be taking greatly increased in April 1982, when the Army released news of the return to training men and women in separate companies. In June, Thurman announced that the panel was "taking a hard look" at changing the 1977 policy that allowed women closer to the forward areas of the battlefield. The chairwoman of DACOWITS wrote Secretary of Defense Weinberger that the delays in the WITA report were "having a very definite effect on the morale of our military women."[93]

In August 1982, the WITA study created even more controversy and concern about repeated delays in release of the final report. Brig. Gen. Ronald Zeltman, who replaced Maj. Gen. Wetzel as director of the study, indicated he planned to classify the study's findings. Classification would mean little or no distribution. On 26 August 1982, the secretary of Defense announced the decision, based on the study group's unreleased findings, to increase the number of enlisted women in the Army during the next five years from 65,000 to 70,000. This figure was well below the goal of 87,500 set by the Carter administration. Simultaneously, the Army announced that the WITA review panel had clarified the definition of *combat exclusion*, making necessary the closing of an additional twenty-three MOSs to women, raising the total to 61, or 52 percent of all Army jobs. (In the course of the study group's deliberations, the Army had accepted the Department of Defense's definition of *close combat* and extended its language to define *direct combat*.[94]) Women in the closed MOSs would have to choose different jobs at the time of

---

[92] Hooker, p. 91; Stiehm, pp. 60, 146. [Hooker in Carol Wekesser and Matthew Polesetsky, eds., *Women in the Military: Current Controversies*, p. 91.]

[93] As quoted by Stiehm, p. 62.

[94] WITA Policy Review, 12 Nov 1982, pp. 7–8.

first-term reenlistment. Revised physical strength standards and tests could be expected to close even more MOSs or to bar women from many specialties.[95]

Meanwhile, DACOWITS took a firm stand. As noted, the advisory group had not been informed of the Army's plans to once again separate the sexes in Army basic training. In addition, at its spring conference in 1981, the group had approved a recommendation that a member of DACOWITS be invited to participate in an advisory capacity on the WITA Policy Review Group.[96] Although Army Chief of Staff Meyer enlisted several DACOWITS' members as advisors to the study group, they were never kept informed and often not invited to meetings.[97] In September 1981, General Thurman held a news conference to announce that the report would be released on 30 September 1981. In late October, it was announced that final preparation of the report was taking longer than expected. DACOWITS had by then begun to press the issue with the Army and the Department of Defense in the same aggressive manner used during its successful attempt in 1967 to pass PL 90-130, which had removed restrictions that prevented women from becoming generals or admirals. DACOWITS' justification on that occasion for what proved to be a major confrontation with the services was the expected increase of the maximum career potential for new recruits. By November 1982, the relationship between DACOWITS and the Army concerning the WITA study had become severely strained.

The WITA Policy Review Group finally released the study during the DACOWITS fall conference at Fort Bragg in mid-November. The three-day meeting was accompanied by substantial press coverage and public statements from the floor about women's issues by enlisted women stationed at Fort Bragg and by general officers who appeared at the Pentagon to explain the report and the Army's plans. During these "stormy and confrontational" sessions, members of the DACOWITS panel and other proponents of equal rights for women in the services accused the Army of having produced "nothing but a snow job" and of poor management. An Air Force officer charged that "the Army developed its conclusions [about women's roles] and then

---

[95] Army Historical Summary, FY 1982, p. 59. For a personal account of the effect of the closing of MOSs, see Lt Col Anna M. Young, USA, "Army Women: Looking Toward an Uncertain Future (Again)," a speech delivered during Federal Women's Week, 26 Aug 1983, Presidio of San Francisco, Calif., published in *Minerva* Quarterly Report on Women and the Military, spring 1984.

[96] DACOWITS History of Recommendations, spring 1981.

[97] Mitchell, pp. 111–22, on DACOWITS' activities.

began looking for a rationale to support them."[98] General Thurman, the DCSPER, and William D. Clark, deputy assistant secretary of the Army for Manpower and Reserve Affairs, among other senior officers, assured the detractors that the report suffered not from mismanagement but from poor public relations.[99] However, the authors of the report saw it somewhat differently: "While this review was needed, in retrospect, its conduct under the aegis of "Women in the Army" was unnecessary. Problems identified resulted from inadequate planning and research as the Army assimilated increasing numbers of women into the force; not because women were in the Army."[100]

As previously noted, the WITA Policy Review Group had begun its work in May 1981 to study the impact of nineteen issues concerning enlisted women, including women recruits in basic training. Also as noted, a year and a half later, only three of those issues remained on the agenda. The question of how pregnant women should be regarded was quickly handed over to the Department of Defense for a solution. That left the Army with two programs to implement, each of which was almost as controversial as the report itself and each sure to have an impact on enlistment and basic training. In designing the new programs, the Soldier Support Center, as lead agency, relied on the data that TRADOC had collected at the Army schools (see above) and on a subsequent TRADOC procedures manual, *Accessing the Physical Demands and Direct Combat Probability of United States Army Operations, Military Occupational Specialties and Duty Position*, of January 1982.[101]

The first of these new programs, known as the Military Enlistment Physical Strength Capacity Test, or MEPSCAT, concerned physical strength requirements. The program would replace the existing practice of randomly assigning recruits to an MOS or of granting individual requests. Using the TRADOC data and the aforementioned TRADOC methodology, the review panel constructed a method of matching the physical abilities of recruits, regardless of sex, to the physical requirements of each job specialty. Each MOS was categorized using four physical tasks: lift, carry, push, and pull. Each task for a particular MOS was measured according to the extent of its physical demands—

---

[98] Quotes are from Mitchell quoting the *Army Times* and Stiehm quoting the *Washington Post*, 9 Nov 1982.

[99] Mitchell, p. 118.

[100] WITA Policy Review report, 1982, p. 1.

[101] Msg, Maj Gen John B. Blount, TRADOC chief of staff to distr, 18 Aug 1981, sub: Implementation of TRADOC Methodology for Assessing What Women Should Do and Where They Should Serve in the Army.

light, medium, and so forth. A test would then be given to potential recruits to determine their abilities to meet the physical requirements of their chosen or assigned MOSs. The MEPSCAT included a measurement for body fat, a hand-strength test, a lifting test, and a test for cardiovascular condition.[102] General Thurman explained that a new "gender-free" physical strength requirement would benefit Army women: "What we [the Army leadership] are trying to do is match the person with the job, both mentally and physically, so they will have a more even-handed chance to stay with [the Army] and get promoted."[103]

Despite Thurman's apparent confidence, leaders concerned with recruiting actions had strongly opposed the new program, even before the 1982 WITA study was released, on the grounds that it would make recruiting goals even more difficult to achieve.[104]

The second program to evolve from the WITA study also used data that TRADOC had collected from commanders in the field. The program—called Direct Combat Probability Coding (DCPC)—examined every MOS and assigned to it a code number ranging from 1 to 7, with 1 being those positions most likely to place the occupant in direct combat. Program managers considered such factors as unit mission, battlefield location and direct exposure to fire, the probability of direct physical contact, and Army doctrine. For example, soldiers serving in combat specialties, position code P1, had to be routinely located forward of the brigade rear boundary. More than half the Army's jobs, 302,000 of 572,000, were rated P1 and, therefore, were closed to women. The closures, however, affected only about 17 percent of MOSs. The DCPC policy replaced the Combat Exclusion Policy of 1977. It also required that position coding be updated at least once per year. It should again be noted that MOS training did not take place during BCT but during AIT. However, enlistees chose the MOS during enlistment, and a decade later, the DCPC would play a significant role in the changes for women in Army basic training following Operation DESERT SHIELD and Operation DESERT STORM.[105]

---

[102] Ibid., pp. 2-1–2-35; Mitchell, p. 109.

[103] Kathryn M. Hodges, "Women in the Army Policy Review"—An Exclusive Interview with Lt Gen Maxwell R. Thurman, vice chief of staff of the Army," *Minerva*, 22 Jun 1983, pp. 86–89. Thurman became vice chief of staff of the Army on 23 Jun 1983.

[104] Mitchell, p. 110.

[105] WITA Policy Review report, 1982, pp. 6–11; Army Annual Historical Summary, 1983, pp. 60–61.

Conclusion

With the advent of the Reagan administration, many military women and women's rights advocates feared that an anti-woman atmosphere or campaign might be expressed as a backlash against the progress made during the 1970s. In general, that reaction was not seen. Although the pendulum took a mild swing toward concern for combat effectiveness and away from social equality issues, there was no major rollback of changes already made. Indeed, with the directives of Secretary of Defense Caspar Weinberger and Assistant Secretary of Defense for Manpower and Reserve Affairs Lawrence Korb, the status quo was maintained and some changes were made, if slowly.[106] Generally, the question of how many women should serve gave way to how the combat exclusion laws ought to be applied in determining how women would serve. By 1982, it was widely assumed that women would never be combatants. But, except for the issue of abortion, the Reagan administration never opposed feminists in any significant way. Even DACOWITS continued to represent the interests of military women after most of its members had been appointed by the Reagan administration.[107]

There was, however, one exception to the absence of a backlash with the advent of the Reagan presidency. The one major military policy change for the Army in the early 1980s was the service's return to the separation of men and women trainees in its basic training program at the entry level.[108] With pressure coming from inside and outside the Carter administration, the Army had established mixed-gender training for enlistees without a clear statement of goals, policies, or procedures. The service seemingly had to determine at every step how the Department of Defense, Congress, the White House, and the public would react to the coeducational training of recruits. Nor was there a clear understanding of how the Army's ground combat role differed from the missions of the Navy and the Air Force, especially with regard to women in combat. The Marine Corps never publicly considered mixed-gender training or women in combat.

---

[106] Lawrence Korb, Ph.D., was especially vigilant concerning the activities of the military services with regard to policies toward women. After leaving the Defense Department, Korb joined the Brookings Institution and became a strong supporter of increasing the number of women in the armed services.

[107] Mitchell, pp. 115, 126. DACOWITS members served three-year terms.

[108] The Air Force, like the Army, reduced the number of women from the goals set by the Carter administration, but no major changes in policy regarding specialties were made.

The entire issue of mixed-gender training in BCT was also a counter-point to the common wisdom that, following the Vietnam War, the Army's leadership had achieved a "training revolution." It was held that the Army succeeded in turning training around, as demonstrated by the Army's performance in Operation DESERT SHIELD and Operation DESERT STORM in the early 1990s. The training of men and women together at the basic level was not a part of the program considered so successful at all levels. Ironically, it took the Persian Gulf War to convince the Army to once again attempt mixed-gender training at the entry level.

The Army's return to gender separation of recruits has been called by some critics of the action "retrenchment" or "abandonment." It must be remembered, however, that the Army's action did not represent a complete turnaround. Because the company is the basic unit for BCT and there would be no more mixed companies, men and women would be physically separated. But the BIET program of instruction adopted in 1977 remained in effect for both genders. In short, the Army had successfully made the transition from separate POIs during the years of the WAC to an integrated POI. In addition, women drill instructors continued to train male recruits.

And studies continued to determine the best use of women in the Army and in the other military services. But it would be twelve years before the service again attempted mixed-gender basic training.

III

An Uneasy Interlude: 1982–1996
● ●

Introduction

During the remainder of the Reagan administration, the first Bush presidency, and the first Clinton term, the Army continued the attempt to define its policy regarding mixed-gender training at the basic level. From the time the Army cancelled the first experiment with such training, in 1982, until after the Persian Gulf War, there was enough action on the subject to maintain some public interest. The presence of women soldiers during operations in Grenada, Panama, the Persian Gulf, and later, in Somalia and Haiti kept the issue of women in combat current in the press and, often, in Congress. As several students of the subject have observed, the general discussion of women in the military tended to shift from the questions of numbers or percentages to how or whether women should serve in operational roles and how the long-standing combat exclusion laws should be applied. It was still generally assumed that women would not be combatants, but definitions of the term *combat* varied widely.[1]

Also contributing to the visibility of the question of women's roles were efforts within and outside the Air Force and the Navy to determine the future of women as combat aviators and as personnel aboard ships with combat missions. This same question of how women should serve was also a central dilemma for the Army. If women were not to be allowed in combat units, could the service justify a "train as you will fight" approach for both men and women in basic training? Indeed, the decision about how women should be utilized would affect training at all levels. The Army debate about mixed-gender training, however, tended to be more internal than that of the other services, depending once again on the time-honored, if not proven, "studies" approach. Some critical observers maintained that studies served as "vehicles for institutional

---

[1] Holm, p. 396.

inertia"; that is, as long as a sensitive issue was being studied, no action was necessary. Others believed that radical steps should not be undertaken without a long and careful analysis of risks versus gain.

The question of combat exclusion for women in the military services came to a head immediately following the build-up in the Persian Gulf and the battles of Operation DESERT SHIELD and Operation DESERT STORM in the early 1990s. Following these operations, which saw the largest deployment of military women in U.S. history to date, the Clinton administration—primarily Secretary of Defense Les Aspin—began to push for the opening of many more job opportunities for women. By 1994, women were allowed service in combat aviation and aboard combat ships. Meanwhile, the Army, realizing that it would not escape this new orientation, had begun planning once again for basic training in mixed companies, which commenced in October 1994 at Fort Jackson, S.C., and Fort Leonard Wood, Mo. At this point, mixed-gender training in basic combat training (BCT) began to be commonly referred to as *gender-integrated training*, or GIT.

The Army's new mixed-gender training for enlistees was, on balance, a successful effort, though some problems and opposition remained, and debate continued regarding the same issues that had surrounded the question of mixed-gender training since the advent of the all-volunteer force (AVF) (see Chapter I). The new BCT program, however, was seriously threatened two years later when female trainees at Aberdeen Proving Ground in Maryland and other locations accused their male trainers of sexual assault and rape. Greatly increased media attention to Army matters and the resulting public outcry once again raised serious questions concerning mixed-gender training at the entry level.

MEPSCAT and DCPC

As discussed in Chapter II, only two programs evolved from a controversial Women in the Army (WITA) study of 1981–1982. The fate of either or both programs had the potential to influence decisions regarding the future training of both male and female recruits. Even before publication of the WITA Policy Review report in 1982, the Army prepared to implement the Military Entrance Physical Strength Capacity Test (MEPSCAT) program and Direct Combat Probability Coding (DCPC), a system designed to determine the likelihood of a unit's mission or a military occupational specialty (MOS) to place a soldier in direct combat (Chapter II). Both these efforts forced decisions regarding how women would serve and, consequently, how they would be trained in BCT. The MEPSCAT matched physical abilities to the physical requirements of each Army MOS. Job specialties fell into five categories of physical demand, and tests occurred at Fort Jackson in the fall of 1982 during pre-basic and the final week of BCT.[2] The U.S. Army Research Institute for the Behavioral and Social Sciences (ARI) validated the tests on 1,003 female soldiers and 980 male soldiers before they began basic training.[3] A total of 64 percent of the jobs were judged to require heavy labor; 42 percent of women were in heavy-duty specialties, but only 8 percent could meet the established standards. The program was formally implemented in 1984 during medical screening, amidst much controversy among Army recruiters, who insisted that if such tests were required, the number of men and women in basic training would be severely reduced. When at least one study indicated that nearly 50 percent of women failed to complete their first enlistment, recruiting became even more of a concern. Others insisted that MEPSCAT would virtually eliminate women from the most "promotable" MOSs. At the same time, company-grade commanders of integrated units voiced concerns that women were still being assigned to positions without regard to their abilities to satisfy strength, stamina, and operational requirements.

For these conflicting reasons, few if any women were restricted from any positions because of physical requirements. Army recruiters prevailed, and testing was done on an advisory basis alone, leaving acceptability decisions to recruiters, who were hard pressed to meet quotas. As one recruiter noted: "...[W]e don't even get involved. The same test is given regardless of the MOS... [and] in two years I've never had a recommendation for a rejection yet. The bottom line is, if they have the minimum smarts and can pass the physical, I sign them up. That's what I get paid for."[4]

Thus, the Training and Doctrine Command (TRADOC) began referring to MEPSCAT as simply a counseling tool. In mid-1983, Secretary of the Army John O. Marsh Jr., conscious of recruitment goals, announced that MEPSCAT tests would be used only as guidelines in the assignment of MOSs.[5]

---

[2] *Performance on Selected Candidate Screening Test Procedures Before and After Army Basic and Advanced Individual Training,* U.S. Army Research Institute of Environmental Medicine, Natick, Mass., Jun 1985. The tests were also conducted at the close of advanced individual training (AIT).

[3] David C. Myers, Deborah L. Gebhardt, Carolyn E. Crump, and Edwin A. Fleishman, *Validation of the Military Entrance Physical Strength Capacity Test,* Technical Report 610 (Alexandria, Va.: ARI, Jan 1984).

[4] Hooker, p. 90.

[5] Mitchell, p. 41.

Not everyone was critical of the physical testing program. Chief of Staff of the Army Gen. Edward C. Meyer, before implementation of MEPSCAT and shortly before his retirement, defended the need for such a program before the Senate Appropriations Committee:

> Previously, the Army had a personnel system which virtually ensured women soldiers would fail. We were ordered to take in a whole bunch of women, and we put them in jobs where they really had no opportunity to succeed. There was never a [female] platoon in basic training that won "Best Platoon." So they felt ill of themselves. The males felt they were not pulling their fair share of the load when we put them all together. From a physical point of view, we put women into jobs which they weren't able to carry out. The men thought the women were not doing their job, so we had harassment occurring.[6]

The DCPC assigned a code number to every position and unit in the Army according to the likelihood that the MOS or the unit would be involved in direct combat. Women were not assigned to MOSs or to units coded P1. No statutory guidelines such as those that pertained to the Navy and the Air Force applied to the Army and the MOS coding program, which the Army established in January 1983 in an attempt to define the intent of the laws that affected the other services. While MOSs were not trained in BCT, the coding of occupational specialties, along with MEPSCAT, influenced the options open to women enlistees and the makeup of the BCT program of instruction.

From the beginning, field implementation of DCPC presented serious difficulties for the Army leadership. It was a personnel management nightmare. Units and MOSs were improperly closed, and women were removed from units contrary to policy. Some women were denied participation in training exercises. As a result, on 15 April 1983, the Army Deputy Chief of Staff for Personnel (DCSPER) suspended the list of closed units and directed the validation of DCPC coding. The revalidated list was approved by the secretary of the Army on 20 October 1983. Meanwhile, the Army, responding to a recommendation by the Defense Advisory Committee on Women in the Services (DACOWITS), transmitted to the field detailed instructions concerning the career transitions of women affected by closed units and MOSs.[7]

---

[6] Hearing before the Senate Appropriations Committee, Subcommittee on Defense on the fiscal year 1984 Army Budget Overview, 98th Cong., 2d sess., 24 Feb 1983 (exchange between Sen William Proxmire and Gen Meyer).

[7] Army Annual Historical Summary, 1983, pp. 61–62; DACOWITS fall conference, 1983.

As with many issues concerning women in the military, DCPC stirred controversy. To DACOWITS and to some women in the Army, DCPC was less a means of excluding women from combat as a scheme to put a ceiling on the numbers of women in the Army. Opponents insisted that women were being removed from units and jobs in which they were performing successfully. Although it was generally agreed that the average young woman was weaker in the upper body than men, was that difference a basis for combat exclusion? Others had difficulty with the various definitions of *direct combat*. Amidst all the uproar, with revalidated data collection and with some pressure from the Department of Defense, the Army restored 13 of the MOSs that had been closed as a result of the original DCPC program.[8] Nevertheless, DCPC continued to erode when P1-coded vacancies went unfilled. In 1987, the Army inspector general declared DCPC to be "unworkable" and recommended that the policy be abandoned. The Army chief of staff chose not to take that route, but in the next couple of years, many more jobs were opened to women.[9]

The 1980s Controversies

As the Army leadership sought to make physical capability and the likelihood of serving in direct combat factors in the assignment of MOSs to recruits, a number of follow-on and related issues and events would influence how women would serve and, therefore, how they should be trained. Once again, the Army turned to studies for guidance. The Army's collective experience of the mid- to late-1980s indicated that although a few changes had been made, the concerns that had prevailed since 1973 remained. The rest of this chapter provides a summary of the gender-related debates and activities surrounding the Army during the twelve years between the first experiment with mixed-gender training in BCT and the second experience. It was against this background that the Army moved once again in the mid-1990s to train men and women together in the same units.

The NORC Surveys

It is difficult if not impossible to separate the questions about mixed-gender training at any level from those concerning women in combat or even the broader topic of women in the military. In April

---

[8] These 23 MOSs represented 52 percent of all Army jobs. Friedl, p. 101.
[9] Holm, pp. 403–07.

1983, the National Opinion Research Center (NORC) at the University of Chicago published an extensive nationwide survey of attitudes concerning women in the military. A surprising 84 percent of respondents supported the idea of maintaining or increasing the number of women serving in the military services, while 81 percent believed that the increased presence of women had not reduced effectiveness or readiness. Only 35 percent expressed a willingness to have women hold direct combat jobs, while there was overwhelming support for women in traditional roles, such as nurses, and in some nontraditional occupational specialties, such as military truck mechanics and fighter pilots.[10] During this time, the Army steadfastly clung to separate training companies in BCT.

Letters (DACOWITS and Weinberger)

The account of DACOWITS' role in the final release and publication of the WITA study of 1981–1982 was described in Chapter II. In the months following the stormy meeting at Fort Bragg, N.C., in November 1982, DACOWITS and its allies continued to attack the WITA report. They criticized as too old to be relevant a Department of Labor categorization of jobs that the Army had adopted to set up the MEPSCAT program. One critic complained that creation of a "moderately heavy" job category amounted to "statistical sorcery." DACOWITS continued to insist that long-range weapons, which made close contact with the enemy a chapter in the Army's history, should not be a part of current operational concepts. Its members once again maintained that the WITA study group had only *assumed* that female soldiers who lacked the designated physical strength for a job downgraded unit readiness.[11]

In April 1983, at its semiannual meeting, DACOWITS recommended "the Secretary of the Army establish a Blue Ribbon Panel reporting to the Secretary, headed by a retired Office of Personnel Management Directorate woman general officer, and composed of active duty/retired Army officers, and noncommissioned officers (with a predominance of women members)."[12] The panel was to review the WITA study and oversee the validation of the review and related studies. On 3 June 1983, the DACOWITS executive board sent a letter to Secretary of Defense Caspar Weinberger stating that the ongoing DCPC deprived the Army of manpower and skilled soldiers and adversely

---

[10] Carolyn Becraft, "Facts About Women in the Military," Women's Research and Education Institute, Jun 1990.

[11] Mitchell, p. 119.

[12] DACOWITS History of Recommendations, spring conference 1983.

affected morale, enlistments, and promotions. The letter asked for a modern definition of *combat*, one that would reflect an increasingly fluid battlefield. Finally, the DACOWITS letter questioned the motives of the review group and the Army's credibility when it came to women: "We have serious questions regarding the merit of the continual studying [of] women's participation. As a study reaffirms the positive performance and contribution by those of our gender, a new one seems to be ordered. This finally raises the question of whether objectivity or "right answers" is the purpose."[13]

On 19 July 1983, Defense Secretary Weinberger sent a memorandum to the service secretaries noting that the press releases regarding the WITA study had left the impression that the Department of Defense had changed its policy toward women in the service. He insisted that combat exclusion be interpreted to allow all possible career opportunities for women. Weinberger's reply to DACOWITS, dated 27 July 1983, was congenial. He assured DACOWITS members and friends that women would be given every opportunity to reach their potential within the limits set by combat exclusion statutes and related policies. As far as the Pentagon was concerned, combat exclusion should reflect the preference of the American people and be decided by elected representatives in Congress.[14]

The Army continued to defend the WITA study until the summer of 1983, when Gen. John A. Wickham Jr. replaced Gen. Edward C. Meyer as Army chief of staff. At approximately the same time, a new assistant secretary of the Army for Manpower and Reserve Affairs, Delbert L. Spurlock, who had had no experience with the WITA study, took office.[15] After talking with Weinberger and Assistant Secretary of Defense Lawrence Korb, Spurlock recommended to Gen. Wickham that the WITA study be reevaluated and errors of methodology be corrected. In October 1983, the Army briefed DACOWITS on the changes that would eventually mean reopening some MOSs, adjusting DCPC, and rectifying the ineffectiveness of the MEPSCAT program (see above).[16]

---

[13] As quoted in Mitchell, p. 120. See also Holm, p. 403. The letter was signed by DACOWITS Chairwoman Mary Evelyn Blagg Huey, president of Texas Woman's University.

[14] Mitchell, pp. 121, 126–27. It may be remembered that there was no statute pertaining to Army decisions regarding combat exclusion, which were made by the secretary of the Army.

[15] Spurlock was a civil rights lawyer with no military experience except two years as Army general counsel. Mitchell, p. 121.

[16] Ibid., p. 122.

Mixed-Gender Basic Training

The policy confusion that characterized the 1980s was indicative of the difficulties the Army faced as it became increasingly clear that the future would likely mean greater gender integration. The service could not expect wholehearted support from the Department of Defense, which usually preferred to hand most problems concerning the training and utilization of women off to Congress. And it was clear that the Army's "train as you will fight" concept would likely force some major decisions concerning men and women in basic combat training.

Grenada, 1983

In October 1983, the question of how the Army should deal with women—at all levels—surfaced in a rather dramatic manner when the first gender-integrated Army units were deployed to Grenada in Operation URGENT FURY. On that occasion, Army paratroopers, marines, and special operations forces invaded the island of Grenada to rescue U.S. citizens and overthrow the Marxist government. In all, 170 female soldiers (Army) served in the operation, most from Fort Bragg. The women served as military police, communications and maintenance personnel, helicopter pilots and crew chiefs, intelligence specialists, stevedores, and medical personnel. Female Army aircrew members flew OH-58 helicopters on operational missions into hostile territory.[17]

During the operation, the service of women in integrated units received some national press coverage, but in general, Defense Department and Army public relations officials attempted to restrict news coverage, in part to avoid alarming the public about the utilization of women as near-combat troops.[18] The Army also had command and control problems, reflecting confusion over its new DCPC system, discussed above. Four women military police officers who were deployed from Fort Bragg were sent back to their barracks three times before being allowed to depart for Grenada. When they arrived on the island four days after the initial invasion, they were ordered back to Fort Bragg while the Army tried to sort out what the policy actually was. The commanding general of the 82d Airborne overruled the action, and the women returned to Grenada.[19]

Debates concerning the role of women in the Army that had plagued the service since the establishment of the AVF in 1973 returned dur-

---

[17] Holm, pp. 404–05, 431.

[18] Enloe, p. 85.

[19] Jennet Conant, John Barry, Verne Smith, Linda Prout, Debbie Seward, and Liz Balmaseda, "Women in Combat: Withdrawing Them in a Crisis Could Hamper Readiness," *Newsweek*, Vol. 106, No. 20, 11 Nov 1985, pp. 36–38.

ing the Grenada operations. Detractors, especially among senior Army retirees, claimed that the participation of women was no proof that they were an asset to unit readiness. To them, women's contributions had been exaggerated by a liberal press and feminists. They insisted that the Army leadership knew the truth but was being muzzled.[20]

DACOWITS and the Risk Rule

In September 1987, as a result of continuing concerns raised by DACOWITS about the full integration of women in the armed forces and, specifically, about basic training in the Army and the Marine Corps, Secretary of Defense Weinberger established the Department of Defense Task Force on Women in the Military. DACOWITS' concerns also arose from observations of sexual harassment and debasement of women during a duty tour of Navy and Marine Corps installations in the Pacific Rim. Although the Pentagon classified the advisory group's report, it was rather widely circulated among the national media. Weinberger directed the task force to address a number of topics affecting women's careers, utilization, morale, and quality of life. Perhaps more important, the task force was to address a perceived inconsistency in the application of combat exclusion statutes and policy across the services.[21]

In late January 1988, Secretary Weinberger's successor, Frank Carlucci, received the report, which made a number of recommendations, including reaffirmation of the Department of Defense policy against harassment and suggestions for even more studies, surveys, and reviews for publication and dissemination. In addition, each service was to review its education and training concerning sexual harassment. The study further recommended that the service secretaries "develop a comprehensive plan to integrate nontraditional skill areas with [sic] enlisted women, with explicit focus on recruiting and assignment policies."[22]

The most important and influential result of the January 1988 Carlucci report was the development of a *risk rule* that set a single standard for evaluating noncombat positions and units from which the services could exclude women. The risk rule permitted closure of noncombatant positions or units "if their risks of exposure to direct combat, hostile fire, or capture are equal to or greater than the risks for land,

---

[20] Holm, p. 440.
[21] Department of Defense, "Report of the Task Force on Women in the Military," Jan 1988, pp. i–ii; Linda Bird Francke, *Ground Zero: The Gender Wars in the Military* (New York: Simon & Schuster, 1997), pp. 29–30.
[22] DoD Task Force, Jan 1988, passim.

air, or sea combat units with which they are associated in a theater of operations."[23]

This new assignment policy was intended to prevent situations where similar positions or units were opened to women in one service but closed to them in another. The new definition did not supersede the DCPC, but it substituted relative risk for location on the battlefield as the key factor. The immediate result of the new combat identification rule was to open an additional 30,000 noncombat job specialties to women as the services reviewed positions previously closed to women. Most of the positions were in the Army. This reexamination, in turn, changed the mixture of MOSs in BCT. The risk rule, however, left the definition of *combat* as obtuse as before and proved equally difficult to implement.[24]

More Studies—GAO

When the Army once again designed and implemented BCT mixed-gender units in the mid-1990s, it would have a multitude of studies as guidance. However, there was seldom agreement with the results of those studies, even among the Army's senior leadership. In July 1988, the General Accounting Office (GAO) completed a study of Senate Bill 581 (S. 581) at the request of Sen. William S. Cohen (R-Maine) and others. The proposed legislation would direct the secretary of the Army to "provide for more efficient utilization of female members of the Army by permitting the permanent assignment of such members to all units of the Army that have as their mission the direct support of combat units." That action—based on function rather than risk—would have the effect of opening all combat support and combat-service support MOSs and units to women regardless of risk. The GAO report concluded that the impact of S. 581 could not be determined until all units had been evaluated under the risk rule.[25]

Two months later, Cohen and Sen. William Proxmire (D-Wisc.), joined by Sen. Dennis W. DeConcini (D-Ariz.), once again asked the GAO to investigate whether the military services might be unnecessarily limiting job opportunities for women. Specifically, they wished

---

[23] Secretary of Defense Frank Carlucci, Memo for Secretaries of the Military Departments, sub: Women in the Military, 2 Feb 1968.

[24] Holm, p. 433.

[25] GAO, "Women in the Military: Impact of Proposed Legislation to Open More Combat Support Positions and Units to Women," Jul 1988. Senator Proxmire was author of the bill. Senator Cohen later served as secretary of Defense in the Clinton administration.

to know what effect combat exclusion actually had on the numbers of women and their assignments and what other policies might have the effect of limiting opportunities. The GAO report found that of 2.2 million military jobs, half were closed to women, but of the remaining 1.1 million open to men and women, not all were being made available to women. How the limits were imposed varied from service to service.[26]

According to the GAO, the Army's accession goals for enlisted personnel were gender specific, thereby limiting the number of women recruited and the number of jobs made available to them. The Department of Defense insisted that the Army's accession objectives for women were "not limits but goals." The Army's Recruiting Command told the GAO that "recruiting is directed primarily toward men, and female accession goals have been met without special effort." The GAO recommended that the number of unrestricted noncombat jobs and the availability of qualified women should govern the maximum number of enlisted women, rather than annual accession goals written specifically for women by the DCSPER.[27]

ARI and Recruitment

The Army's training of men and women in mixed companies in the future depended, at least in part, on whether or not women would be recruited for combat roles. Paramount for recruiters was the question of whether removing or modifying combat exclusion laws and policies would have any impact on the nature and effectiveness of recruitment policies and programs. In early 1990, when an unsuccessful bill was introduced in Congress that would have required the Army to recruit women for combat for a four-year trial period, the Army directed the Army Research Institute to investigate the potential need to recruit women and to study public attitudes and perceptions of enlisting women in Infantry, Armor, Artillery, and Combat Engineers branches.[28]

ARI researchers, using the results of five public opinion polls and attitude surveys, found that approval of recruiting women into the combat arms had increased during the 1980s. There was, however, still no consensus and heated debate continued. Before men and women could choose from all of the Army's MOSs and be successfully integrated in

---

[26] GAO, "Women in the Military: More Military Jobs Can Be Opened Under Current Statutes," Sep 1988.

[27] Ibid., quotation, p. 4.

[28] Mary Sue Hays and Charles G. Middlestead, "Women in Combat: An Overview of the Implications for Recruiting" (Alexandria, Va.: ARI, Jul 1990), pp. vii–ix. In 1990, 214 of the Army's 258 MOSs were open to women.

basic training, the prospect of women in combat would have to be promoted to the public. The role of women as combatants would change from one of being trained in a noncombat specialty that included contingency training in general combat skills to one of being recruited, trained, and evaluated as dedicated combat soldiers. The Army already knew of demographic trends that seemed to indicate that the available pool of new recruits would become increasingly more female; if that eventuality occurred, the Army's Recruiting Command would have to address the attitudes of potential enlistees, as well. In sum, the existence of a policy that permitted women in combat would dictate the need for an effective and extensive public relations effort.[29]

The Field Artillery Question

The Field Artillery branch provided a clear example of the "combat conundrum." Since the late 1970s, a number of Field Artillery MOSs had been closed to women, then reopened, only to be closed again, primarily because the term *combat* had proved difficult if not impossible to define. In February 1978, Secretary of the Army Clifford L. Alexander Jr. opened Field Artillery, with the exception of Cannon Field Artillery, to women. In November 1982, the DCPC system permitted women to serve in headquarters and service batteries but not in firing batteries. In September 1988, Army Chief of Staff Carl E. Vuono approved closing Field Artillery to women to avert career progression problems caused by the demise of the Pershing missile system and the drawdown of the Lance missile system.[30] Upon the request of the DCSPER, the CSA left the Lance batteries open until a follow-on system—presumably the Multiple Launch Rocket System—was ready. The rocket system would be forward deployed and closed to women, a situation that effectively closed career progress for women in the Field Artillery. In November 1988, for that reason and based on a TRADOC report titled *Review of Positions Closed to Women*, the TRADOC training commander, Lt. Gen. John S. Crosby, recommended closing Field Artillery to women, an action that was taken in December 1989. DACOWITS took a strong stand against such action and once again insisted that women should be allowed in any position for which they were qualified. The committee asked for a four-year trial period. In June 1989, a compromise recommendation was accepted by the Army chief of staff, leaving the Field

---

[29] Ibid., pp. v, 2, 7, 25.
[30] The Lance was a short-range missile designed for use against the former Soviet Union. It was capable of carrying nuclear warheads. The system was decommissioned in 1991–1994.

Artillery branch open to women temporarily but providing for no additional Field Artillery positions in the future.[31]

All these controversies and issues were interrelated and directly affected the training of men and women in basic training. Each time debate intensified as to the wisdom of mixed-gender training for recruits, especially in the Army, each of these issues resurfaced in the Department of Defense, among Army leadership, in Congress, and in the academic and public press. For example, discussion of training men and women together at the basic level was sure to raise questions about women in the draft and further questions about women in combat. What would be the effect of such training on military readiness and the physical conditioning of the troops? Against this background, the U.S. Army made a second attempt to establish gender-integrated basic combat training.

Panama and Operations DESERT SHIELD and DESERT STORM

Beginning in the last days of 1989 and continuing through the fall of 1991, three events changed the way the Army looked at its sexually segregated training system for enlistees. The first event was the invasion of Panama in late 1989. Next came Operation DESERT SHIELD and Operation DESERT STORM in the Persian Gulf in 1990 and 1991. Finally, in September 1991, came the Navy's Tailhook sexual misconduct scandal.

In the early hours of 20 December 1989, the U.S. military undertook combat operations in Panama to oust dictator Manuel Noriega. In all, 770 women served in Operation JUST CAUSE. Women participated in noncombat positions as military police and in supply, communications, and transport operations. Army women also flew Black Hawk helicopters ferrying infantry troops to landing zones, often under heavy fire. Circumstances forced at least two women to command troops in combat when they encountered enemy soldiers. The lines between combat and noncombat duty had become, at minimum, blurred.[32]

Major stories appeared in almost all national news publications regarding women's service in Panama. Unlike Grenada, this time, the press and broadcast media had not been kept at such a distance. Reports brought the increasing importance of women in the military to the attention of the public, many of whom still believed that women

---

[31] Fact Sheet, ATCD-SP, 30 Oct 1989, sub: Women in Lance Battalions; TRADOC Annual Historical Review (AHR), 1989, p. 47.

[32] Holm, pp. 405, 431, 434–35.

were excluded from combat. Military analyst Charles Moskos termed it "a shot heard around the world, or at least in the Pentagon."[33] The Army, surprised by the level of public interest, initially had little to say. Spokesmen usually insisted that the women's participation had been overemphasized in the press. While some senior military officials praised the women's accomplishments, others insisted that Operation JUST CAUSE offered no proof that women were an asset to unit performance.[34] A brief flurry of legislative activity followed the operations in Panama, but by March 1990, when the House of Representatives Military Personnel and Compensation Subcommittee of the Committee on Armed Services held its post-Panama hearings on women in the military, television networks and popular news publications had lost interest.[35]

Five months later, however, women in the military once again became a front-page story. On 2 August 1990, the Iraqi army crossed the border of Kuwait and overwhelmed the Kuwaiti forces, an action that threatened to destabilize world oil markets and the entire Middle East. During the next week, Iraqi forces moved dangerously close to the Kuwaiti–Saudi Arabian border. As war threatened in the Persian Gulf area, President George H. W. Bush ordered Operation DESERT SHIELD, the deployment of ground, air, and naval forces to the region. Operation DESERT SHIELD was the first major deployment since Vietnam and the largest deployment of women in U.S. history, to date.

DESERT SHIELD and the combat operations that defeated the Iraqis in early 1991 (DESERT STORM) saw a total of 40,782 women deployed, 30,855 of whom were Army personnel. Female soldiers made up 9.7 percent of the total number deployed. Thirteen women were killed, and two women were taken prisoner.[36] Women pilots from the 101st Airborne Division delivered supplies and personnel. Women soldiers directed artillery, operated prisoner-of-war camps, maintained tanks, and worked as truck drivers, cargo handlers, intelligence and communications specialists, paratroopers, and flight controllers.[37]

---

[33] *Washington Post*, "Women in Combat: The Same Risks as Men?" 3 Feb 1990.

[34] Holm, p. 435; Robin Rogers, "Combat Exclusion Promotes Widespread Discrimination in Society," in Carol Wekesser and Matthew Polesetsky, eds., *Women in the Military: Current Controversies*, p. 128.

[35] Enloe, pp. 88–89

[36] Holm, pp. 439, 469; Friedl, p. 101. Approximately 540,000 military officers and enlisted personnel were deployed to the Persian Gulf. Eleven percent of active-duty military were women; 13 percent were reservists. Deployment figures tend to vary. See GAO, "Women in the Military: Deployment in the Persian Gulf War," 13 Jul 1993.

[37] Melinda Beck, Ray Wilkinson, Bill Turque, and Clara Bingham, "Our Women in the Desert" *Newsweek*, Vol. 116, No. 11, 10 Sep 1990, pp. 22–25.

In the 1990–1991 Persian Gulf War, it was clear from the beginning that there were no distinct "lines in the sand." The front was constantly changing, and noncombat units were often as exposed to attacks from surface-to-surface missiles, as were those on the front lines. The DCPC system continued to cause confusion within the Army and greatly complicated the management of Army personnel. Necessity often forced military commanders to assign soldiers without regard to gender.[38] The broadcast media and the press, experiencing almost total Pentagon and White House control of news coverage from the Gulf, turned to human interest stories about military women. That approach meant that women assumed news value disproportionate to their relative numbers. From the beginning, there had been considerable concern in the Pentagon that having women in combat would be a traumatic experience for the American people. However, despite the extensive press coverage, the public generally accepted the presence of women in the theater of operations. As expected, some skeptics, inside and outside the Pentagon, contended that the war had been too short and the enemy too poorly equipped to provide a test of integrated forces. And some male military personnel deeply resented the press's implication that they were more expendable than females.[39]

By the time the shooting war started in the Persian Gulf in January 1991, representatives of the National Organization for Women (NOW) and DACOWITS were filling congressional hearing rooms as senators and representatives offered a plethora of bills for and against women's role in war. On 8 May 1991, Rep. Patricia Schroeder (D-Colo.), of the House Armed Services Committee, chaired by Rep. Les Aspin (D-Wisc.), in a "stunning political coup" attached repeal of the 1948 combat exclusion law against women pilots to the 1992 Defense Appropriations Bill. The vote was unanimous for repeal. This action transferred the authority to assign women to combat aircraft from Congress to the service secretaries. On 22 May 1991, the Defense Appropriations Bill passed the full House.

The repeal legislation had greater difficulty in the Senate, beginning with its first hearing on 18 June 1991. The amendment went before the Senate Armed Services Committee, chaired by Sen. Sam Nunn (D-Ga.). Initially, Sen. John Warner (R-Va.) and Sen. John McCain (R-Ariz.) of the committee supported the measure, only to withdraw their support a month later. At that point, Senator Nunn and Sen. John Glenn (D-Ohio)

---

[38] Holm, pp. 446–47.

[39] Ibid., pp. 439–41, 470–71; Stephanie Gutmann, *A Kinder, Gentler Military: Can America's Gender-Neutral Fighting Force Still Win Wars?* (New York: Scribner, Mar 2000), p. 136.

maneuvered to have the amendment replaced by a proposal for a presidential commission to study the role of women in the military. To have a commission study an issue and make recommendations took pressure for a decision off Congress and the Pentagon. To counter that strategy, Sen. Edward M. Kennedy (D-Mass.) and Sen. William Roth (R-Del.) proposed joint legislation to bypass the committee and take the issue directly to the floor of the Senate. That tactic set off a lobbying campaign led by representatives of NOW, DACOWITS, the American Civil Liberties Union, and the Women's Military Aviators organization, to name a few. By conducting what one observer termed "a blitzkrieg" in the Senate corridors, the women's rights groups aimed to persuade the Senate into following the House in voting for repeal of the combat exclusion law. Intense lobbying against repeal was led by Phyllis Schlafly, president and founder of the Eagle Forum, and Elaine Donnelly, president of the Center for Military Readiness. On 25 July 1991, Senator Kennedy and Senator Roth announced joint legislation. A week later, a voice vote held in the Senate approved repeal of the legal statute against women pilots, but it also approved the establishment of a presidential commission. At that point, Secretary of Defense Dick Cheney announced that no women would be assigned to combat aircraft until after submission of the commission report, due in November 1992.[40]

As the American public expressed a renewed appreciation for the armed forces after the operations in the Persian Gulf and as repeal of the combat exclusion law for women aviators worked its way toward implementation, an event involving Navy pilots affected the way all the services approached gender issues. During the Labor Day weekend on 5–7 September 1991, the naval aviators' Tailhook Association lost Navy sponsorship when attendees at the annual convention were implicated in incidents of alcohol abuse, destruction of property, and sexual misconduct and assault. Senator McCain denounced the Navy's handling of the affair on the floor of the Senate, and the Department of Defense inspector general called for an independent investigation. Meanwhile, the news media kept before the public an image of male-dominated military services whose senior officials tolerated sexual harassment and discrimination. As a representative of the GAO would later observe, "after Tailhook, everything was about gender."[41]

---

[40] Francke, pp. 220–33; Holm, pp. 488–92; Mitchell, p. 214; Gutmann, p. 148. The repeal of the combat exclusion law for women aviators was signed in Dec 1991 but did not take effect until after the commission report in Nov 1992. The remainder of the combat exclusion law remained in effect.

[41] Gutmann, pp. 157–58; GAO, quotation, p. 157; Enloe, pp. 92–93; for a detailed but biased account of the Tailhook scandal, see Mitchell, pp. 260–69.

The Presidential Commission on the Assignment of Women in the Armed Forces

Beginning in March 1992, the Army, especially those at TRADOC responsible for BCT, carefully watched the proceedings of the Presidential Commission on the Assignment of Women in the Armed Forces. As noted, the commission grew out of the participation of women in Panama in 1989 and Operation DESERT SHIELD and Operation DESERT STORM in 1990–1991 and the resulting legislative battles over repeal of the combat exclusion law. The commission focused on the assignment of women to combat aircraft and to combatant ships but also considered recommendations regarding physical qualification standards, affirmative action, cohesion in mixed-gender units, and the principles under which military personnel policies should be established— all issues important to the question of training men and women in the same companies during BCT.[42]

The Presidential Commission was made up of fifteen members: nine men and six women. Among the commissioners was Gen. Maxwell R. Thurman (Ret.), who had served as the Army's chief of recruiting; DCSPER; vice chief of staff of the Army; commanding general of TRADOC; and commander in chief of the U.S. Southern Command during Operation JUST CAUSE. Also on the commission was Elaine Donnelly, president and founder of the Center for Military Readiness, a former member of DACOWITS, and a conservative lobbyist opposed to mixed-gender training for Army enlistees. The commission was headed by Gen. Robert T. Herres USAF (Ret.), former vice chairman of the Joint Chiefs of Staff. Immediately, the commissioners chosen by President George H. W. Bush and Secretary of Defense Cheney were assailed as too conservative by those favoring greater choice for women in the military services, but a careful reading of the commissioners' biographies does not seem to bear out that accusation.[43] From March to November 1992, the commission held thirty-two public hearings and traveled extensively in the United States and abroad, at a cost of $4.1 million. The commission was to decide what recommendations to make regarding whether "existing laws and policies governing the assignment of [military] women...should be retained, modified, or repealed."[44]

---

[42] Unless otherwise noted, all information on the Presidential Commission on the Assignment of Women in the Armed Forces is from the commission's published report to the president, 15 Nov 1992.

[43] Francke, p. 240.

[44] Quotation from the commission's report, p. iv.

The commission's sessions were turbulent and divisive. Debates were emotional and highly charged and offered no easy answers. Arguments tended to revolve around the question of military readiness. Advocates of greater opportunities for women stood by their argument that the deciding factor should be qualification, not gender. Readiness, they insisted, would be enhanced by a larger pool of applicants. Advocates of the male status quo held that quotas would be unavoidable, thereby allowing less-qualified women into key positions and compromising readiness.[45] Opinion polls showed military personnel opposed to a change in the status quo; the public was about equally divided on the issue. Meanwhile, as the commission began to vote on its recommendations, five of the most conservative members walked out and were not present for the vote. The commission adopted its recommendations the same day that William J. Clinton defeated George H. W. Bush for the presidency. The recommendations proved, as expected, to be controversial and contradictory. The commission voted seven to six against women being assigned to combat aircraft until further tests were conducted. In a seemingly contradictory action, the law banning women from combat vessels, except amphibious ships and submarines, was repealed when General Herres appealed to the members to show a willingness to give something to the reformers. Otherwise, he argued, "people will not believe we credibly considered the issues."[46] The Army's senior leadership breathed a sigh of relief when the commission recommended that women continue to be banned from Infantry, Armor, Combat Engineer, special operations forces, and some artillery MOSs.

This, then, was the military, social, and political background against which the Army would make a second attempt at mixed-gender training at the basic level. Although there were considerable misgivings and, in some cases, outright opposition among the Army's senior leadership, the choices were dwindling. The 1990s had begun with what one writer called "an exhilarating explosion of activism."[47] Bill Clinton, the newly elected democratic president, was certain to hold more liberal views on the utilization of women in the military than his predecessor, George H. W. Bush. Clinton also would bring a democratic Congress into office

---

[45] Capt Mimi Finch, USA, "Women in Combat: One Commissioner Reports," *Minerva*, spring 1994, p. 2; Francke, pp. 246–55.

[46] Enloe, p. 104. While the commission met, American women began to take part in military operations in Somalia (Nov 1992–Mar 1994) as the United States participated in United Nations efforts to establish a secure environment for humanitarian relief operations. In all, more than 1,000 women performed support roles in Somalia.

[47] Davis, p. 491. The Clinton victory brought an unprecedented number of women to Congress, and of Clinton's first 500 political appointments, 37 percent were women.

with him. The Presidential Commission was ongoing, but no one was taking bets on its outcome. In addition, by the fall of 1992, one in every seven cadets at the U.S. Military Academy at West Point was female. On 1 March 1990, the publicly funded Virginia Military Institute (VMI) in Lexington, Va., was sued by the Department of Justice on the grounds that the military college's exclusion of women violated the equal protection clause of the Fourteenth Amendment to the Constitution. In the fall of 1992, the suit was still pending before the Circuit Court after initial rulings had been in VMI's favor, but the new administration would undoubtedly continue appeals.[48]

Fort Jackson, ARI, and Secretary Les Aspin

Collectively, these events brought grave concern to TRADOC's training managers. The political climate and the Clinton administration's support for women's programs seemed to indicate that some action on integrated training needed to take place if the Army was to compete in the recruitment wars. But the Army encountered a no-win situation. The Department of Defense claimed that Congress and the American people should decide the issue of combat exclusion, not the Pentagon or the services. Further, if the Army adopted policies aimed at broadening opportunities for women, traditionalists blamed the service for compromising readiness and catering to women's rights advocates; if a policy took anything away, DACOWITS, NOW, the news media, and the Office of the Secretary of Defense charged the Army with being discriminatory and regressive. In such an atmosphere, decisions were often more political than military.

Supporters of mixing men and women in basic training included U.S. Army War College commandant Maj. Gen. Richard A. Chilcoat, who briefed Army Chief of Staff Gordon R. Sullivan on the subject. Chilcoat and others argued that times had changed and that women currently served with men in all of the Army's noncombat positions. If one of the Army's foremost principles was to train soldiers as they were going to fight and support, did it make sense to train men and women separately during their first eight weeks in the service?[49] Advocates for

---

[48] In Oct 1992, the case was still being argued in the Fourth Circuit Court of Appeals. In 1995, President Clinton instructed his administration to file a brief asking the Supreme Court to declare that government actions that discriminate on the basis of sex should be subject to the same strict constitutional scrutiny that the Court applies to official distinctions on the basis of race. On 26 Jun 1996, the Court ruled that VMI must either forego state funding or admit women. Thomson Gale, Free Resources.

[49] BCT was expanded to nine weeks in 1998.

the program, doubtless attempting to head off concern over the major criticisms of earlier attempts, pointed out that at least some testing had shown that mixed training did little to affect the physical conditioning, marksmanship, or individual proficiency scores of men but did cause a striking increase in the morale and performance of women.[50]

Finally, in January 1993, after months of considering the pros and cons, TRADOC commander Gen. Frederick W. Franks Jr. directed Fort Jackson to test gender integration of basic combat training companies down to squad level. The use of Fort Jackson as a testing center mirrored actions taken in 1976–1977, when testing of the mixed-gender BCT program first took place (see Chapter II). Fort Jackson conducted a three-phase test from March to November 1993, during which performance data was assessed to determine whether the integrated model affected measurable performance indicators. In all three phases, data was collected on basic rifle marksmanship, Army physical fitness training, individual proficiency tests, sick call, profiles (exclusion from training for medical reasons), and graduation rates. The Phase I and II tests involved gender integration of two companies for two consecutive training cycles, one having a male-to-female ratio of three to one; the other company organized with a one-to-one ratio. Four gender-pure companies served as controls for the tests. Phase III involved integrating six companies, four with a three-to-one ratio (male-to-female) and the other two each with a one-to-one ratio.[51]

The trainers with the 1st BCT Brigade at Fort Jackson concluded that "the soldierization and training performance of the gender integrated company was equal to, or better than, the soldierization and training performance of the four gender-pure companies."[52]

They also encountered no significant problems in integrating male and female soldiers through the squad level. The key to success, they concluded, was the attitude of the cadre, an issue that would repeatedly surface as the Army sought the best approach to integrating men and women in small units.[53]

If officially adopted, the new BCT program, unlike the earlier one, would feature completely sex-integrated training, even coed barracks.

---

[50] Bradley Graham, "In Coed Training, Army Revisits a Basic Strategy," *Washington Post*, 21 Nov 1994, pp. A-1, A-10.

[51] Annual Historical Summary, Fort Jackson, 1993, pp. 99–100; Memorandum, 1st BCT Brigade for commander, U.S. Army Training Center and Fort Jackson, 19 Apr 1993, sub: Gender-Integration Test After Action Report. Phase III was designed to identify the best "mix" of male and female soldiers for basic training.

[52] Ibid.

[53] Ibid.

Like the earlier program, the program of instruction would be the same for both sexes. Male and female recruits trained on the same courses, shot the same rifles, and carried weighty gear. However, the physical performance requirements differed on the grounds that men had larger hearts and lungs, more muscle mass, and longer strides. Men had to be capable of performing 32 push-ups and 42 sit-ups and of running two miles in 17 minutes to receive an average score. The standard for women was 13 push-ups, 40 sit-ups, and two miles in 20 minutes. Ability groups for running were formed early in each training cycle in an attempt to have trainees compete against those at their level, with achievement measured by the amount of improvement. Abandoning the "gender-free" requirements of the MEPSCAT effort (see Chapter II), the aim was to attain the same level of expenditure of energy by men and women.[54]

In an effort to avoid many of the problems that plagued the earlier integrated training program, in June 1993, General Franks directed ARI to participate in Phase III of the training tests at Fort Jackson and to measure and analyze trainee and cadre attitudes toward what had now been dubbed universally as GIT.[55] The "attitude assessment," conducted in the fall of 1993, focused on three areas of concern: soldierization (defined as the process that turns civilians into soldiers); soldier attitudes toward BCT and the Army; and the training cadres' attitudes toward gender-integrated BCT.[56] The ARI assessment team chose a focus group methodology designed to draw out more answers than the standard "pick-one" questionnaire so widely used in military research. To produce the necessary data, pre-training and post-training soldier surveys and post-training cadre surveys were developed in close coordination with Fort Jackson. Each participant in the focus groups answered the same battery of questions, for example: What were the

---

[54] Graham. Late in 1997, the physical training requirements were brought more in line for men and women.

[55] Unless otherwise noted, all information in this section is from the draft of the report, titled "Gender Integration of Initial Entry Basic Combat Training." This report was apparently never published. TRADOC Regulation 350-6 establishes the gender-integration policy for Army BCT. In accordance with this policy, in 1993, BCT was integrated at battalion level but conducted in single-gender companies. TR 350-6 was revised in Nov 1998, Jul 2001, and Aug 2003. As of late 2004, 350-6 was once again being revised to include lessons learned from Iraq (see Chapter V).

[56] A second phase of the testing at Fort Jackson took place at Fort Leonard Wood, the only other site where mixed-gender training would initially take place, in the summer of 1994. The tests at Fort Leonard Wood included one battalion of four companies, all integrated with 75 percent males and 25 percent females. The author was unable to locate any records at Fort Leonard Wood.

best and worst things that happened during training? Were there differences in drill sergeants' treatment of male and female soldiers? Other data was obtained through observation of the training exercises.

In two consecutive training cycles during the testing at Fort Jackson, two training battalions of five companies each formed the sample. In each battalion, one company was all-male, one all-female, two were integrated at the three-to-one ratio, and the fifth company was integrated at a one-to-one ratio. Focus group discussions were conducted with all-male or all-female groups, each made up of eight soldiers from all of the training companies (a total of 128 trainees) and with male and female members of the training cadre.

Some of the study results were surprising, and probably disappointing, to the Army leadership who were determined to make the new GIT program work, even if many of them were opposed to it in principle. Others worried about the possible effect on recruitment if the new program failed. The assessment team concluded that the ratio of women to men did not affect soldiers' commitment to the Army nor their physical or mental growth. On the other hand, individual morale tended to be highest for males in single-gender units, lowest for females in single-gender units, and relatively balanced in integrated units. Levels of unit cohesion—a major concern among the Army leadership—were low for all trainees but highest among males in nonintegrated units and lowest for females in those units. Levels of teamwork were lowest for females in single-gender units and improved among women in integrated units.

ARI found soldier attitudes toward the Army were positive throughout BCT regardless of gender. Attitudes toward BCT were most positive for males in single-sex companies and least positive for females in single-sex companies, suggesting that expectations or the actual training experience was different for female recruits than for male recruits. Female interviewees in integrated units believed they were challenged more and pushed themselves harder than females in nonintegrated companies. Many of the male trainees believed they were expected to achieve more, work harder, and perform better while females were not expected to live up to the same standards. Most females thought men and women were treated about equally. Many males found the training not challenging enough, a situation that those in gender-integrated units blamed on the presence of women. Conversely, those in nonintegrated companies blamed the lack of rigor on low entrance standards. Some female interviewees complained that the drill sergeants were too easy on them, making training "more like a summer camp."[57] Most male

---

[57] ARI Report, 1997, p. 20.

soldiers exhibited positive attitudes toward women in the Army prior to training than after BCT, regardless of the type of unit. Both men and women rated their drill sergeants higher on technical skills and lower in "people" skills.

For their part, the training cadre complained about the poor physical condition of trainees upon arrival, the inadequate number of drill sergeants, and the length of the duty day. Combat arms drill sergeants were not as supportive of mixed-gender training as those in combat support and combat service support. The cadre in combat arms was more likely to be critical about the benefits of gender-integrated training for men or women and the ability of female soldiers to meet the physical demands of BCT. A majority of the cadre expressed a preference for training non-integrated units. A number of trainees complained that drill sergeants with negative attitudes adversely affected their experience and attitude. Not surprisingly, the ARI assessment team concluded that "training of cadre is the key to successfully implementing any changes."[58] However, noncommissioned officers overwhelmingly believed gender-integrated training to be a mistake. As one commented, "Where did the machine break that it needs to be fixed?"[59]

In general, the survey team found that training was equally as effective in mixed companies as in segregated companies, with a preferred ratio of 75 percent men to 25 percent women. As one of the members of the attitude assessment team put it, the men in 75/25 units, compared to 50/50 units, "felt they were still in control."[60] Researchers also found that gender integration had no effect on graduation rates or career intentions. Physical training and sick-call rates tended to improve in mixed units, as did morale, cohesion, and teamwork for females. ARI personnel stressed that TRADOC needed to design a new "fill" plan to support mixed-gender training, and plans were needed for modifications to some barracks. Regardless of the mix of males and females, ARI recommended that classes in personal hygiene and rape prevention remain single gender.

The Army leadership believed the ARI survey, on balance, revealed successful gender integration of BCT. But the program still had its critics, inside and outside the Army, some of whom had been consistently critical since the early experiments in training young enlistees in mixed-gender units. The issues these critics dealt with in the 1990s were remarkably similar to the concerns that had existed at least since

---

[58] Annual Historical Summary, Fort Jackson, 1993, p. 99.
[59] ARI Report, 1997, p. 49.
[60] Graham, p. 4.

the inception of the AVF, such as physical strength, stamina, and sexual harassment.

As soon as the new pilot programs began, the print and broadcast media, remembering Operation DESERT SHIELD and Operation DESERT STORM and the ongoing debates about women in combat, began to report on the Army's new effort to establish a mixed-gender program for BCT. In addition, several full-length studies were published—most but not all of which were critical. Some observers worried that mixing the genders in basic training would open the door to allowing women in combat MOSs. Perhaps the most frequent negative observations addressed the differing physical standards for men and women. Such arrangements as ability groups, critics charged, were not only unfair to men but gave an advantage to women. Adjustment of the fitness criteria meant that men were not challenged and women would not fail. In any case, ability grouping often meant resegregation of the sexes, a situation that hurt cohesion and morale. In internal Army publications, commanders often complained that female recruits suffered more injuries and illnesses, thereby slowing down the training schedule. Other critics were concerned that gender-mixed training would mean "pull[ing] the men down to the women's level."[61]

Indicative of the controversies that swirled around the issue of training men and women in mixed companies at the basic level were two books published in the early years of the 1994 renewal of mixed-gender training. The first was *A Kinder, Gentler Military: Can America's Gender-Neutral Fighting Force Still Win Wars?* by Stephanie Gutmann, and the second was *Weak Link: The Feminization of the American Military* by Brian Mitchell. Both authors were vehemently opposed to the change in BCT. Gutmann, for example, accused the Army of having a "de facto gag rule" concerning the subject of gender integration: "Everyone knew someone who'd been fired or penalized in some way for saying something incorrect about the way the integration project was going. Women had become 'the third rail'; not wanting to risk saying 'the wrong thing,' commanders who were having troubles with their new mixed-sex units simply shut their mouths or used approved language with extraordinary care."[62]

Mitchell, who served as a witness before the Presidential Commission on the Assignment of Women in the Armed Forces, earlier addressed the issue of fraternization, a major concern for the Army leadership and training cadre: "Nothing has done more to cheapen rank and diminish

---

[61] Ibid., quotation, p. A-10; Francke, p. 37.
[62] Gutmann, pp. 16–17.

respect for authority than cute little lieutenants and privates."[63] The list of critics is long, but a common theme was that the military existed to defend the nation, not to engage in social engineering.

Problems or no problems, critics or no critics, several events in 1993 and 1994 limited the Army's choices with regard to mixed-gender BCT. Further consideration of policy changes or execution regarding women in combat, specifically the prohibition on women flying combat missions, depended on the recommendations of the commission, which filed its report on 15 November 1992. Five months later, the new Clinton administration's Secretary of Defense Les Aspin rejected the commission's recommendation that the status quo be maintained. In his Memorandum on the Assignment of Women, dated 28 April 1993, Aspin directed the military services to open more occupational specialties to women, specifically assignment to aircraft engaged in combat missions. The Army moved quickly to code positions in attack helicopter battalions as "interchangeable."[64] Aspin also directed the Navy to draft legislation to allow women to be assigned to combat ships, except submarines and the Navy's Sea, Air, Land forces. The Army and the Marine Corps were to study opening new assignments to women.[65] Justification was required for any assignments that remained closed to women.

On 13 January, Secretary Aspin made an even bolder decision to rescind the risk rule that had governed what assignments women could hold since 1988. The rule, Aspin believed, had become inappropriate. In the future only occupational specialties involving direct ground combat would be restricted. *Direct ground combat* was defined as "engaging an enemy on the ground with individual or crew-served weapons, while being exposed to hostile fire and to a high probability of direct physical contact with the hostile force's personnel."[66] The new definition, which would still prohibit women in Armor, Infantry, Combat Engineer, Cannon Field Artillery, forward-area Air Defense Artillery, and special

---

[63] Brian Mitchell, *Weak Link: The Feminization of the American Military* (Washington, D.C.: Regnery Publishing, 1989), p. 176.

[64] Msg, Cdr Personnel Command to distr, 061350Z, May 1993, sub: Women in Attack Aircraft.

[65] ATCD-SE to Cdr TRADOC, 1 Mar 1994, sub: Assignment Policy for Women in the Army. Secretary Aspin was especially concerned that Field Artillery and Air Defense Artillery positions be studied for opening to women.

[66] Memo, Secretary of Defense to Secretaries of the Army, Navy, Air Force; Chairman, Joint Chiefs of Staff; Assistant Secretary of Defense for Personnel and Readiness; and Assistant Secretary of Defense for Manpower and Reserve Affairs, 13 Jan 1994, sub: Direct Ground Combat and Assignment Rule.

operations, was to go into effect on 1 October 1994.[67] This change in direction opened 32,000 Army jobs to women.[68]

The climate of the Clinton administration was significantly different from that which had prevailed in the early 1980s, when the Army abandoned its first attempt to integrate training of both genders at the basic level. If the new gender-integrated training program in BCT was not a success, this time, the Army would have fewer choices. The new laws pertaining to the Navy and the Air Force cut the statutory ground from under the Army's policy and, the service worried, might threaten recruitment. On 17 August 1994, Army Chief of Staff Gen. Gordon R. Sullivan announced the total integration of Army training, to take effect on 1 October 1994, the same day that the risk rule would cease to have effect.[69] "Train as you will fight" had taken on added meaning.

Fort Jackson and ARI Revisited, 1995

Two months after the concept of direct ground combat replaced the risk rule in determining what jobs women could hold, the new TRADOC commander, Gen. William W. Hartzog, established a GIT Steering Committee to make recommendations on whether, and how, training policies should be altered to ensure the successful long-term implementation of mixed-gender BCT. Subsequently, ARI was directed to design a study that could provide information to the committee. This study is often referred to as the ARI Phase III study, the first two having taken place at Fort Jackson in 1993 and Fort Leonard Wood in 1994. The focus of the Phase III study was similar to that of the first two. ARI looked at trainees' attitudes and their evaluation of their training experiences, drill sergeants' attitudes and evaluation of the drill sergeant course, suggestions for improvement, attrition of soldiers during BCT, the physical condition of soldiers at entry through graduation, and the training performance of soldiers.

The study was once again conducted at Fort Jackson and Fort Leonard Wood. Four training companies from Fort Leonard Wood and six from Fort Jackson were included. All were gender-integrated with fills that varied from 23 percent to 48 percent female. Data collec-

---

[67] The first women were assigned to Navy combat ships in Nov 1993. The first women flew combat aircraft in Apr 1994.

[68] While more jobs were opened to women, it was a common complaint that often, these new "gender-free" positions were not being filled with women.

[69] Navy and Air Force chiefs generally supported the new legislation, as did the Army's civilian leadership, Congress, and the general public. The Marine Corps, at this writing, still does not train men and women together in basic training.

tion began in April 1995 and continued through September 1995. The data was provided to the Steering Committee in December 1995.[70] The methodology was almost identical to that of the first two studies, that is, pre- and post-training questionnaires and focus groups. Data from the study was compared, where possible, with that of the 1993 and 1994 studies, which had single-gender companies.[71]

Much of the study focused on the entry-level physical condition of male and female soldiers. The study team found that few soldiers physically prepared themselves to enter BCT and most entered in poor condition, females more so than males. While drill sergeants were usually able to bring most recruits up to standard, the physical toll on the soldiers and the time required by the drill sergeants for remedial physical conditioning was a detriment to training overall. The lack of a physical fitness standard for accession required the screening and separation of soldiers unable to meet the physical demands of BCT or the Army.

In addition, first-term attrition was most often caused by poor conditioning, resulting in injuries. The study team recommended a physical fitness standard for accession at the recruiting centers.[72] Implementation of MEPSCAT in 1982 had been intended to provide such screening but failed. When recruitment began to lag, MEPSCAT lost emphasis. In any case, the study report was quick to point out, improvement had occurred "without the Army fitness standards being changed or adjusted for gender-integrated training." The statement countered some of the more vocal critics of mixed-gender training in BCT, who had charged that a change in standards, rather than actual improvement, was allowing females to compete. Just as previous studies had concluded, men and women in integrated companies performed as well if not better than those in single-gender companies, except that men in single-gender companies performed better in the push-up events. Some males continued to complain that the women slowed them down. However, well-conditioned males in all-male companies had the same complaint about less-fit males.

---

[70] In Jan 1996, data from the study was provided to the GIT Steering Committee and to the assistant secretary of the Army for Manpower and Reserve Affairs. The data was reviewed by a GAO report to the chairman of the Subcommittee on Military Personnel, Committee on National Security, House of Representatives in Jun 1996. The study was published in Feb 1997.

[71] Jacqueline A. Mottern, et al., "The 1995 Gender Integration of Basic Combat Training Study" (Alexandria, Va.: U.S. Army Research Institute for the Behavioral Sciences, Feb 1997).

[72] To pass the Army physical fitness test, a recruit in BCT had to score 50 points on each of the push-up, sit-up, and two-mile run tests. The Army standard was 60 points. For the remainder of their careers, soldiers would have to meet the Army standard.

To move from BCT to AIT, soldiers required extensive training in the use and maintenance of the M-16 rifle. A soldier had to score at least 23 of a possible 40 hits on electronically scored targets to qualify in basic rifle marksmanship. Although males tended to qualify more often on the first attempt, by the end of the marksmanship training day, there were no significant differences in qualification between male and female trainees. Successful completion of another series of tests, termed the Individual Proficiency Tests, showed similar results. The proficiency tests were designed to test a trainee's achievement in twenty skills or abilities in which soldiers received instruction during BCT. These generally addressed the ability to perform first-aid, the wearing of the mission-oriented protective posture protective mask, firing of the AT-4 light antitank weapon, and use of battle zero (a line-of-sight setting that enables a firearm to shoot on target) on the M-16 A1 rifle. These common skills were graded by noncommissioned officers not assigned to the training company. The ARI study team reported that there were no significant differences, regardless of the degree of female integration.[73]

Of special importance to the Army's leadership and to the ARI team was the measure of "soldierization," which was determined by a composite of self-reported levels of pride, commitment, individual improvement during BCT, morale, teamwork, and cohesion. Generally, there appeared to be no differences between soldiers trained in gender-integrated companies and those trained in single-gender companies. If anything, female soldiers in gender-integrated companies reported higher levels for the components of soldierization compared with those trained in an all-female environment. A number of female trainees indicated that they felt more challenged by training with males and that the drill sergeants tended to push them harder. Levels of teamwork and cohesion improved for females in mixed companies and remained about the same as in previous studies for males in all-male and gender-integrated companies. In private interviews, some male soldiers complained that female soldiers took advantage of their sex with the male drill sergeants. For their part, female recruits complained that males did not show respect for females in leadership positions. Regardless, soldiers of both sexes were very positive about the Army when training began and remained positive at the end of BCT.

Both trainees and their trainers identified gender discrimination and sexual harassment as having occurred during training. Discrimination was reported by 24 percent of the women compared to 7 percent of the men. Drill sergeants and soldiers of both genders reported that some

---

[73] Ibid., p. viii.

male drill sergeants expected less of female soldiers and treated them differently than male soldiers. Female drill sergeants reported that they had received unequal treatment during the drill sergeant course. The study provided soldiers with a definition of *sexual harassment*: "Sexual harassment is a form of sexual discrimination that involves deliberate or repeated unwelcome sexual advances, requests for sexual favors, and other verbal or physical conduct of a sexual nature."[74]

Soldiers were asked if they had experienced such harassment since their enlistment and where this harassment had occurred. Many more females than males reported sexual harassment. Twenty-five percent of females reported incidents of sexual harassment compared to 5 percent of males. The harassment occurred most often in the barracks or during training activities. Other trainees accounted for most of the incidents, with the drill sergeants the second most frequently mentioned source. For about half of those who made reports of sexual harassment to their chain of command, corrective action was taken about half the time. A majority of soldiers who did not report incidents either thought them too trivial or feared retribution.[75]

Perhaps the most significant findings in the 1995 ARI study of gender-integrated training had to do with the attitude and actions of the drill sergeants. Of the drill sergeants serving with the test companies at Fort Jackson, 84 percent were male and most had a combat arms MOS. Most were married. The drill sergeants believed that the drill sergeant course of study did not prepare them to conduct BCT with males and females together. In addition, a majority of drill sergeants did not believe gender-integrated training should be continued. Most said they did not want to train females at all—they preferred to train all-male companies—but would rather train mixed companies than all-female companies. Some of the trainers believed training female soldiers made them afraid to "act naturally" for fear of being charged with using improper language or with sexual harassment. On the other hand, some of the trainers thought that competition between the genders encouraged all soldiers to perform better and reduced fights and bickering between single-gender companies. Great variation existed among the drill sergeants regarding mixed-gender training in noncombat arms, but the common denominator seemed to be the attitudes of the drill sergeants toward female soldiers and toward training in a gender-integrated environment. One of the authors of the ARI report commented: "When drill sergeants continually point out the differences in males and females, tell the soldiers that stan-

---

[74] Army Regulation 600-20, *Army Command Policy*, 30 Mar 1988.
[75] ARI Report, 1995, pp. 31–32.

dards had been lowered for females, do not treat female drill sergeants with respect, and encourage the idea that it is a man's Army, a lack of respect and teamwork among soldiers is fostered."[76]

The ARI study team also looked at the reasons why recruits sometimes left BCT before graduation. Some who were opposed to mixed-gender training at the beginning level—both within and outside the Army—had claimed that training men and women together was a basic cause of premature separation, or attrition. Several prior studies had upheld that assessment. During the 1995 study at Fort Jackson, 142 of a total 1,997 soldiers did not graduate. The available data showed that those who left before completion were less committed to the Army and less confident in their abilities to perform in BCT before starting training. The most common reason for leaving was a pre-existing medical condition, followed by a failure to adapt, injury, personal problems, mental stress, and inability to achieve. There were no differences between graduates and those who separated before graduation in age, race, education, or marital status. However, female trainees, trainees with children, and trainees from communities with populations of more than 25,000 left BTC in higher numbers than expected. When asked what the Army could have done to prevent their attrition, those who had left before graduation generally thought that the recruiters should have provided more information about BCT and the Army prior to the beginning of training. Others believed that reporting for training in good physical condition would have made a difference. In general, the data did not indicate that training men and women together in BCT resulted in increased attrition.[77]

The Reception Station and Beyond

What was the experience of the young men and women whose first eight weeks of life as a soldier were spent in mixed-gender training companies?[78] The following is based not on any particular training cycle but on a composite of several early (1994) GIT experiences. Some components, such as the program of instruction, remained the same when the Army abandoned mixed-gender training in 1982 (see Chapter II), but there had been changes.

When a new recruit joined the Army, he or she was transported to one of the Army's training centers—in the case of gender-integrated

---

[76] ARI Report, 1995, pp. viii, 42, 53.
[77] ARI Report, 1995, pp. ix, 49–50, 55.
[78] A ninth week of BCT was added in the late 1990s (see Chapter IV).

training, Fort Jackson or Fort Leonard Wood—to be in-processed through a reception battalion. Each training battalion was made up of four or five training companies. Each company usually had between 200 and 250 soldiers divided into four platoons, each of which had four squads. In the reception battalion, male and female soldiers were assigned to single-gender platoons and housed in separate barracks. The platoons of fifty to sixty trainees were the basic organizational units for managing the flow of soldiers through the reception battalion, where the new soldiers received a general orientation; completed personnel, medical, and financial records; and received immunizations. To advance to a training battalion, they were also screened for their ability to complete push-ups, thirteen for males and one for females. Recruits who failed to perform the required number of push-ups were assigned to a fitness training company, where they had up to twenty-one days to meet the standard of twenty push-ups for males and six for females. Subsequent failure meant separation from the Army.

Enlistees usually remained at the reception battalion for at least three days. While there, some soldiers received a surprise: Recruiters in many cases had not informed the recruits that they would be trained in gender-integrated units. That situation was upsetting to some recruits and to some parents, most of whom had strongly supported their son's or daughter's decision to join the Army. Not surprisingly, spouses and significant others were also unenthusiastic.[79]

As the new trainees moved from reception battalion to training battalion, some to mixed-gender units, certain demographic statistics emerged. One of five of the enlistees was a woman. Considerably more females than males were African American. Age and education levels were difficult to determine because of the presence of those who had chosen a split option in which high school students could complete BCT in the summer between their junior and senior years. Among males and females, 90 percent were single, and 10 percent had dependent children. A large majority came from towns and suburbs rather than large cities. Most had lived at home with their parents or guardians before joining the Army. Almost all had been previously employed full- or part-time.[80]

One of the major objections to training men and women together in BCT arose from the assignment of barracks for mixed-gender recruits. Depending on the type of barracks, usually a platoon was housed together on a floor or in a bay. In gender-integrated training, it was

---

[79] ARI Report, 1997, pp. 6, 16–17.
[80] Ibid., pp. 10–12.

not always possible to house a platoon on the same floor or bay. In the "starship" style barracks, a bay could house sixty soldiers and had a self-contained latrine.[81] In mixed-gender training, this arrangement meant that females were in one bay with female soldiers from another platoon, while males were in a separate bay, sometimes with soldiers from another platoon. In the "rolling pin" barracks, all females were either located in one section of the floor or on one floor with males on another—typically the second and third floors. Either of these arrangements made it necessary for a drill sergeant to supervise soldiers from a platoon other than his own when his trainees were spread out over two locations. Communications and conflicting orders could become a problem, and platoon cohesion and teamwork suffered.[82]

Other problems with barracks life included rooms with no doors, too few latrines with showers, and disputes over washer/dryer rights. The lack of privacy made it necessary—and officially required—for soldiers to change clothes in the latrine, an inconvenience with a busy training schedule that began at 4:30 a.m. Consequently, most soldiers changed quickly in their rooms. For female soldiers, privacy was an important issue, especially in the showers. For male trainees, privacy was not a major issue. Trainees at Fort Jackson in 1994 frequently mentioned the shortage of washers and dryers, a situation exacerbated by poor quartermaster support. This shortcoming was especially difficult if the machines were located in one latrine on a floor, making it necessary to schedule the opposite gender to go into the latrine to do laundry. As one commander of a training battalion at Fort Leonard Wood put it, "gender-integrated training is no big deal; the big deal is gender-integrated living."[83]

Neither male nor female soldiers in BCT had much personal or free time. For those in single-gender companies, there was no time to socialize with the opposite sex. Those in gender-integrated units had more opportunity, although these encounters were tightly controlled and often depended on command philosophy. Living quarters and latrines were strictly off limits to the opposite sex, and violations might mean loss of pay or restriction to post. Male and female enlistees could converse with each other while on pass, meet during squad meetings, and sit together during church services. Unacceptable behavior included dating or meeting privately with another trainee of the opposite sex, passing personal

---

[81] A starship is a three-story building with five wings, with platoon areas separated by doors that may be secured.

[82] ARI Report, 1993–1994, pp. 28, 46.

[83] "This Woman's Army," *St. Louis Post-Dispatch*, 26 Feb 1995, p. 1F; ARI Report, 1997, pp. 42, 53.

notes, entering the sleeping area of another trainee without permission, and any kind of sexual activity or intimate physical contact. Also prohibited was sexually explicit or obscene language.[84] Fraternization was strictly prohibited by regulation, but the degree of enforcement usually depended on the individual commander. Fraternization occasionally occurred in single-gender and mixed-gender units at sick call, in the laundry room, or in the hallways. More frequent were rumors of female soldiers fraternizing with drill sergeants, by all accounts as exaggerated as they were abundant. Female soldiers were resentful of male soldiers' attitudes that they used sexual favors to advance.[85]

After leaving the reception station, recruits' time became that of the trainers—the drill sergeants who would teach them the basics of soldiering. Trainees were rapidly made to understand that they would be expected to think, look, and act like soldiers at all times.[86] A training company consisted of a company commander, a training officer, a first sergeant, ten to twelve drill sergeants, and support staff. Each platoon had at least two but sometimes as many as four drill sergeants, who were responsible for day-to-day training until graduation eight weeks later. It was the drill sergeants who provided new soldiers with assistance, support, knowledge, and discipline and instilled Army values and behavior. Trainees could expect approximately 85 percent of the drill sergeants to be male. The scarcity of female drill sergeants was a concern throughout the Army. Trainees—male and female—often remarked that every platoon should have at least one female drill sergeant. The male drill sergeants likewise believed that if BCT were to be permanently gender-integrated, it was important to have a female drill sergeant in each platoon. In 1994, there were not enough female drill sergeants to support this staffing. Further, trainees and drill sergeants alike expressed the need to increase the number of drill sergeants to at least three per platoon. These additional trainers should be available full-time for training and not be assigned to other unrelated duties.[87]

The senior trainers at TRADOC were determined to see that training and duties during BCT did not become gender-specific to encourage teamwork and avoid resentment between male and female trainees. Digging of fighting positions was done by all, not just males. Security patrols included men and women, not just women. Enlistees marched

---

[84] ARI Report, 1993–1994, p. 28; *Soldiers*, Department of the Army Public Affairs, Mar 1995, p. 13; Diane Suchetka, "Service Ventures Back into Coed Training: Hanky-Panky? Not in This Army," *New Orleans Times-Picayune*, 4 Dec 1994.

[85] ARI Report, 1993–1994, pp. 28–29.

[86] In accordance with TRADOC Regulation 350-6.

[87] ARI Report, 1993–1994, p. 46; ARI Report 1997, pp. 3, 42.

together, ate together, and threw grenades together. Trainees shared physical training, rifle marksmanship and AT-4 light antitank weapons training, the confidence course, the bayonet assault course, and sessions in the chemical gas training facility.[88]

There were, however, a few instances where men and women were separated. Courses in rape prevention and personal hygiene were taught for women only, at least as late as December 1994.[89] Men and women did not spar against each other in training with the pugil stick (a pole with padded ends used to simulate bayonet fighting). In addition, the BCT "buddy system" assigned each trainee a buddy of the same sex. The buddy system was designed to protect both drill sergeants and soldiers and to teach teamwork. The training cadre indicated that, if enforced equally, the system worked well. However, there appeared to be a double standard, with the buddy system enforced more often for female soldiers and male drill sergeants. The latter were not allowed to talk alone with female soldiers, but female drill sergeants could talk alone with male soldiers, and male officers were permitted to be alone with female soldiers. At least the buddy system solved the problem of sleeping arrangements during field training exercises.[90]

The reaction of the new soldiers to gender-integrated training was as varied as their personalities and experiences. Some believed that the Army was being less than honest in the effort and that the service was not truly dedicated to making GIT work. Some men believed women were a "drag" and that their presence lowered training standards. Others feared some women might out-perform them. Some male recruits thought women brought competition and a challenge to training, which motivated them to work harder. Women trainees feared that men would laugh at their mistakes or go out of their way to make life miserable. Women were especially vulnerable to injuries, and the fierce rivalry too often brought women to hide injuries, resulting in permanent damage. Some drill sergeants, perhaps a majority, considered women a distraction, especially when a third of their time was devoted to controlling male-female relationships. On balance, however, the level of women's performance went up while that of men remained the same.[91] Generally, acceptance was the order of the day. As one training battalion commander said: "The rest of the Army is gender-integrated, so why not

---

[88] Suchetka, *New Orleans Times-Picayune*, 4 Dec 1994; *St. Louis-Post Dispatch*, 26 Feb 1995.

[89] These courses were integrated early in 1995 at Fort Leonard Wood. *Soldiers*, March 1995.

[90] ARI Report, 1993–1994, pp. 5, 26–27, 46.

[91] Ibid., p. 42.

start the process here at the basic training level?...I think the whole group is more cohesive when it is gender-integrated. If the males are in one company, and they see the females in another company, they tend to think the grass is greener on the other side of the hill. Now that they are integrated, they've found that it is not."[92]

Despite some problems still awaiting solutions, the Army leadership, if not a majority of the trainers, were generally satisfied with the second attempt to initiate mixed-gender training at the Army's basic training level. That was the situation until late 1996, when reports of sexual assault and rape in initial-entry training at Aberdeen Proving Ground and other locations caused the Army leadership, the public, the Congress, and the White House to once again question the training of men and women together in BCT.

---

[92] *Soldiers*, Mar 1995.

IV

A New Challenge but Old Issues 1996–2000

•••••••••••••••••••••••••

Aberdeen Proving Ground, November 1996

The United States Army was on a roll in the fall of 1996. As a result of its highly successful performance in the Persian Gulf War, public confidence and pride, esprit de corps, and professionalism were soaring. Recruitment was up, especially among young women as career tracks formerly designated for men opened. In September, women began to be assigned to aviation unit maintenance troops of division cavalry squadrons.[1] Women were performing peacekeeping missions in Bosnia (13,000 women by the year 2000). At Fort Jackson, S.C., and Fort Leonard Wood, Mo., training brigade commanders, their staffs, and drill sergeants continued to work out the day-to-day adjustments of training men and women together in basic combat training (BCT).

Then, an incident at Aberdeen Proving Ground, Md., in advanced individual training (AIT) signaled what appeared to be problems in gender-integrated training throughout the Army. On 7 November 1996, the chief of staff of the Army, Gen. Dennis Reimer, and the commanding general of the Training and Doctrine Command (TRADOC), Gen. William W. Hartzog, reported to the Senate Armed Services Committee and to the public that they had received allegations of sexual harassment and misconduct at the Army Ordnance Center and School.[2] A number of female trainees had accused their drill sergeants of everything from rape to improper touching and condescending remarks. By March 1997, Army investigators had identified 50 victims, interviewed 1,800 witnesses, and suspended 20 instructors. In the following weeks, eleven noncommissioned officers (NCOs) and one captain were charged with

---

[1] Msg, Department of the Army headquarters to distr, 101201Z Sep 1996.
[2] Secretary of the Army Togo D. West Jr. and Army Chief of Staff Dennis J. Reimer to the Senate Armed Services Committee, 4 Feb 1997.

sex crimes under the Uniform Code of Military Justice and Army regulations.[3] Letters of reprimand went out to the Aberdeen commander and three other senior officers. Meanwhile, similar allegations came from Fort Leonard Wood and Fort Jackson. A hastily set up hot line eventually received a total of 8,300 calls, 1,350 of which led to criminal investigations, and 239 of which were still ongoing months later.[4] These indications of a larger, endemic problem brought about a significant public outcry and unwelcome (to the Army) media and congressional attention to the gender-integrated BCT program.

It would be difficult to overstate the tension and threat that the sexual abuse allegations brought to the relatively new Army gender-integrated basic training program. Those inside and outside the Army who had opposed it all along could find plenty of evidence in the situation for returning to gender-pure basic training. Others, ignoring the fact that most of the charges came in AIT, tended to place the blame on the newer BCT program. The sexes had been integrated in AIT for combat support and combat service support specialties since 1975. All subsequent investigations would emphasize the training of men and women together at the basic level as the most contentious and, for some opponents in and out of Congress, the *only* issue.[5]

As the debate about mixed-gender training heated up again, all the old issues came to the fore—pregnancy, unit cohesion, physical fitness, male bonding, fraternization, and so forth. In an attempt to prevent the sexual misconduct allegations from serving as an opening wedge for opponents to claim that military women were not capable of fulfilling their mission, the Army's senior leadership was quick to announce that there would be no rollback in the numbers of women in the service. In a Pentagon press briefing on 6 February 1997, Secretary of Defense William S. Cohen, Army Chief of Staff General Reimer, and Secretary of the Army Togo D. West Jr. told reporters that it was not mixed-gender training that should be blamed but, rather, some leaders' failure to uphold the code of conduct and look after their troops. General Reimer

---

[3] Robert J. Grossman, "It's Not Easy Being Green. . . and Female," *Human Resources,* Sep 2001. These soldiers were also charged with being in violation of TRADOC Regulation 350-4, "Initial Entry Training." Four of those charged were imprisoned; eight were discharged or punished administratively. One drill sergeant received twenty-five years in prison for numerous counts of rape and abuse.

[4] Megan Arney, "Military Report Admits: 'Sexual Harassment Exists Throughout Army,'" *The Militant,* 28 Sep 1997. The hot line was originally established at Aberdeen Proving Ground and was moved to the Pentagon on 12 Dec 1996. Gerry J. Gilmore, "Chief Speaks on Sexual Harassment," *Soldiers,* Mar 1997.

[5] PBS interview with Elaine Donnelly of the antifeminist Center for Military Readiness, 30 Apr 1997.

said: "Sexual harassment is not a policy issue; it's an issue of right and wrong. Sexual harassment and misconduct...lower morale and destroy teamwork and cohesion. They erode respect for the chain of command... and most importantly, they destroy basic human dignity."[6]

The Army Investigates

The Army was quick to confront the situation at Aberdeen, a task made more difficult by a vague definition of *sexual harassment*. The doctrinal underpinnings of the Army's approach to the problem rested in Army Regulation 600-20, *Army Command Policy*, which defined *sexual harassment* as "a form of gender discrimination that involves unwelcome sexual advances, requests for sexual favors, and other verbal or physical conduct of a sexual nature when...such conduct has the purpose or effect of unreasonably interfering with an individual's work performance or creates an intimidating, hostile, or offensive working environment."[7]

Although that definition was in line with the one used in the civilian sector, it was vague enough in the military services to encompass a wide range of behaviors and to allow relationships to quickly become harassment or accepted behavior to become unwanted behavior. The chance of such an outcome was enhanced in the intense environment of basic training.

In addition, as soldiers accused of sexual misconduct were tried in military courts, the concept of *constructive force* (in contrast to *physical force*) was increasingly used by prosecutors. *Constructive force* referred to implied force in situations where one party was in a position of dominance and control. None of these observations should be thought to imply that the Army suddenly discovered sexual misconduct in the fall of 1996. For example, a series of Army Audit Agency reports of 1982 cited widespread perceptions of sexual mistreatment. An Army exit survey of 1985 found that one-third of the women interviewed called sexual harassment a very important reason for leaving the Army.[8] In addition, the designers of the 1994 integrated BCT had insisted on instruction about sexual harassment.

---

[6] Linda D. Kozaryn, "Senators, DoD Leaders Address Sexual Harassment," Armed Forces Information Service, 6 Feb 1997.

[7] AR 600-20, 5 Mar 1993, para. 7–4. This AR was revised on 15 Jul 1999, but the definition of *sexual harassment* did not change.

[8] Army Audit Agency, "Enlisted Women in the Army," Apr 1982; Glenda Y. Nogami, "Army Exit Survey," briefing for Defense Advisory Committee on Women in the Services (DACOWITS), 24 Apr 1985.

Perhaps more important than any of the preceding was a major survey completed in May 1995 by a Department of Defense Equal Opportunity Council Task Force on Discrimination and Sexual Harassment. Co-chaired by Secretary of the Air Force Sheila Widnall and Undersecretary of Defense for Personnel and Readiness Edwin Dorn, the task force held nineteen formal meetings and received a series of briefings from representatives of the military departments, including the reserve components, and subject-matter experts. In addition, they heard the views of advocacy groups and reviewed more than fifty documents, policy papers, and studies. Topics that provided the framework for the survey included service policies, training, complaint-handling procedures, protection from reprisals, complaint appeals, and the responsibilities of leadership. In short, most of the issues identified after the events at Aberdeen had just been examined shortly before the Aberdeen allegations. The task force found that 55 percent of women and 14 percent of men reported receiving "uninvited or unwanted sexual attention" while on duty. Although reports of sexual harassment had "declined significantly" since the last such survey in 1988, when 64 percent of women and 17 percent of men reported such behavior, its incidence was still alarmingly high.[9] The task force concluded that there was a perception that sexual harassment was not taken seriously enough; however, "there is confidence that leaders will deal with it." In August 1995, the report's recommendations were incorporated into the Department of Defense's Military Equal Opportunity Program and its Affirmative Action Planning and Assessment Process. The report did not receive any considerable attention until the sexual abuse charges in November 1996, after which it became required reading for members of the House of Representatives and Senate and those serving in defense agencies.[10]

Congress and the media asked why, given the existence of such a report, the Army was completely surprised by the Aberdeen revelations. In a question-and-answer session with the Department of the Army Public Affairs Office, Secretary of the Army Togo D. West Jr. explained, "we have been relying on statistics that make us feel a little bit better about our handling of the problem than we have a right to rely on. But if you have in place for several years…a procedure that the Congress and others are saying is a model that the other services

---

[9] For an account of the 1988 study, see Chapter III.

[10] Defense Manpower Data Center, Department of Defense 1995 Sexual Harassment Survey, pp. iii–viii. Although Army leaders apparently received copies of the report shortly after its approval, it was not made public until 26 Jun 1996. DefenseLink 26 Jun 1996.

should adopt...[you can be] completely surprised when something like this happens."[11]

Secretary West was the first to react to the allegations of the female trainees. Two weeks after Army Chief of Staff Reimer announced, on 7 November 1996, the charging of five soldiers at Aberdeen with crimes that included rape, fraternization, and sexual harassment and of three NCOs at Fort Leonard Wood with consensual intercourse and indecent assault (touching), West announced two new studies and an action plan. He directed the Department of the Army Inspector General (IG) Lt. Gen. Jared L. Bates to investigate the training base and prepare a document with the descriptive title "Special Inspection of Initial Entry Training, Equal Opportunity/Sexual Harassment Policies and Procedures." The IG was to review sexual harassment training, the complaint process, soldiers' confidence in the system, and the drill sergeant selection and process. He was also to assess whether there were factors in addition to the inherent superior–subordinate relationship that might cause trainees in the training base to be susceptible to sexual abuse.[12]

At the same time, Secretary West announced the membership and charter of a military-civilian task force, the Senior Review Panel on Sexual Harassment, to be chaired by Maj. Gen. Richard S. Siegfried (Ret.), former commander at Fort Jackson and former deputy IG of the Army.[13] Other members of the nine-member panel included Brig. Gen. Evelyn Foote (Ret.), Maj. Gen. Larry Ellis, Maj. Gen. Claudia Kennedy, Deputy Assistant Secretary of the Army for Manpower and Reserve Affairs Sarah Lister, and Sgt. Maj. Eugene McKinney. West directed the new panel to "conduct a comprehensive review of the Army's policies on sexual harassment and the processes currently in place....I am particularly concerned about behaviors that fail to acknowledge the dignity and respect to which every soldier is entitled.[14]

---

[11] 22 Nov 1996.

[12] Army Public Affairs News Release No. 96–82, 22 Nov 1996.

[13] Originally, this investigation was to take place at TRADOC, where the TRADOC IG would report to the commander, Gen William W. Hartzog. It was decided not to have TRADOC do the report because it was necessary to provide "the confidence that it is being done by someone outside of his [Hartzog's] command headquarters." The original announcement of the investigation placed it with TRADOC, leading the media to believe that there was some hidden reason for the action to move to the Department of the Army. Army Public Affairs, transcript of press conference, 22 Nov 1996.

[14] Secretary of the Army, Memo for the Inspector General, sub: Directive for Inspector General, 20 Nov 1996 (quotation), and Memorandum for Maj Gen Richard S. Siegfried, sub: The Secretary of the Army's Senior Review Panel on Sexual Harassment,

The panel was given 120 days to produce an initial report and 45 additional days to submit the final report.[15] West's goal was to use the findings of both the review panel and the IG's investigation to produce a Humans Relations Action Plan. Meanwhile, soldiers worldwide were to receive refresher training in the prevention of sexual harassment.

As the review panel prepared to address the secretary's guidance, the IG arranged to collect the data necessary to his inspections, and the new secretary of Defense, William Cohen (24 January 1997–20 January 2001), made several lengthy trips to the field to assess the situation personally, the Senior Review Panel lost one of its members. Army Sgt. Maj. Eugene McKinney, the service's highest ranking NCO, was himself accused of sexual assault and harassment. His accuser, a retired female sergeant major, did not initially file her complaint with the Army but instead took her story to the *New York Times*. McKinney denied all charges but asked to be excused from his duties on the review panel.[16]

The review panel and the IG's office completed their work in July 1997, but the results were not made public until 11 September 1997, ostensibly to give Secretary West time to respond but more likely to await a new action plan. The panel's efforts resulted in more than 35,000 interviews at 59 bases worldwide. The panel claimed to have found that 47 percent of the Army women polled reported they had experienced sexual harassment. Accompanied by Army Chief of Staff Reimer, Secretary West, in a press conference, outlined the key findings of the two reports and the actions planned to address the issues. "Sexual harassment exists throughout the Army, crossing gender, rank, and racial lines." Drawing a distinction, West asserted that sexual discrimination was more prevalent than was harassment or abuse. Too many leaders, he said, "have failed to gain the trust of their soldiers." He continued by saying that the Army lacked a firm commitment to the equal opportu-

---

21 Nov 1996. Siegfried and Foote were called back to active duty. Siegfried had been commander of the training center at Fort Jackson when gender-integrated training was reestablished in 1994.

[15] Army news release, 22 Nov 1996.

[16] Jamie McIntyre, CNN News, 3 Feb 1997. McKinney was suspended on 10 Feb 1997, and charged on 10 May 1997. Although the incident was not reported until four years later, another panel member was involved in an alleged case of sexual assault while serving on the panel. Maj Gen Claudia Kennedy accused another general officer of sexual assault as he prepared to become the Army's deputy IG, a post in which he would have supervised investigations of sexual harassment complaints. The *New York Times* reported, "Maj Gen Larry Smith requested early retirement after military investigators endorsed sexual harassment charges brought against him by Lt Gen Claudia Kennedy; there were no plans to reduce his rank or retirement benefits through a Grade Determination Review Board." 8 Jul 2000.

nity program and that female soldiers distrusted the complaint process or feared reprisals for reporting abuse. West expressed his belief that respect as an Army core value was not institutionalized in the initial entry training process. The selection, training, and supervision of drill sergeants had to improve. Finally, more pressure needed to be asserted on the chain of command to be responsive to issues of harassment and discrimination.[17] In a surprisingly direct statement, the panel's report concluded that "the human relations environment of the Army is not conducive to engendering dignity and respect among us."[18]

Not surprisingly, reactions to the scathing report were varied. There was widespread praise for the Army's vigorous response, along with condemnation of soldiers' behavior and the command climate that fostered it. House of Representatives Del. Eleanor Holmes Norton (D-D.C.) spoke for the Congressional Caucus on Women's Issues, calling the report "nothing short of refreshing," and its findings "bold and unequivocal." Delegate Norton characterize the report as "pretty stark, pretty frank, and the kind of straight talk that will pierce the ranks up and down….We will be looking to see if they carry out the report with the strength that its language implies."[19]

Rep. Patricia Schroeder (D-Colo.), known for her work on behalf of women's rights and the first woman to serve on the House Armed Services Committee, termed the report "very thorough." On the other hand, Sen. Olympia Snowe (R-Maine) called the report an "indictment of the climate and lack of leadership that permits sexual harassment to permeate all levels of the Army." She added, "I…will continue to aggressively pursue changes to eliminate the poisonous environment that allows such pervasive levels of sexual harassment to undermine the good order and discipline of the United States Army, so crucial to our national security."[20]

Other observers, vocal opponents of the Army's policies on training men and women together in BCT, saw the reason for the self-deprecating nature of the Army's report in the service's memory of the Navy's Tailhook experience (see Chapter III). Brian Mitchell, a former member of the 1992 Presidential Commission on the Assignment of Women

---

[17] "The Secretary of the Army's Senior Review Panel on Sexual Harassment, 11 Sep 1997," passim; Department of the Army Inspector General, "Special Inspection of Initial Entry Training," 22 Jul 1997, passim; Gerry J. Gilmore, "Sexual Harassment Panel Reports Review Findings," Army News Service, 12 Sep 1997.

[18] Senior Review Panel on Sexual Harassment report, p. 2.

[19] House of Representatives, *Congressional Record*, 17 Sep 1997, H.R. 7485.

[20] Senate, *Congressional Record*, 11 Sep 1997, p. S. 9240. Senator Snowe was, however, a strong supporter of gender-integrated training.

in the Armed Forces (see Chapter III), wrote, "the Army shifted into high gear to show that it had learned from the Navy's experience not to take such things lightly."[21] Stephanie Gutmann, who believed that "the complete integration of women into the military is physically and sociologically impossible," wrote of the Army's response to the sexual harassment charges: "Sensing a full-fledged scandal in the making, the Army moved quickly to demonstrate its openness; whatever you do, one could almost hear the brass saying, we are not going to be accused of 'cover up' like the Navy."[22]

It is not entirely clear whether the Navy's experience was a major influence in the Army's reaction to the events at Aberdeen, but on several occasions, Secretary West asked media representatives to remember that "the Army found these conditions and the Army disclosed them."[23]

Secretary West and General Reimer moved quickly to transform the information from the two reports into an action plan designed to put the Army's ethical standards back on track. The plan, formally titled "The Human Dimensions of Combat Readiness," was released in September 1997 alongside the review panel's report. The plan addressed seventeen issues, such as leadership, human relations training, equal opportunity policy, Army core values, and support for the training base. Responsibility for a vast majority of numerous actions and recommendations went to TRADOC, with the remainder shared mostly by the Department of the Army deputy chief of staff for operations, the deputy chief of staff for personnel, and the Department of the Army Personnel Command.[24] West and Reimer explained that the action plan presented a means to "establish an Army environment where soldiers treat one another with dignity and respect...." They defined a successful human relations climate as one that maximized soldiers' awareness of how individual actions could affect their unit's ability to accomplish its mission. The plan also focused on respect among soldiers regardless of race, gender, or ethnic heritage.[25]

To demonstrate the Army's determination to rectify what was seen as a failure of leadership, Reimer introduced or accelerated some programs of his own. "Plain and simple," he said, "this is a leadership

---

[21] Mitchell, p. 309.

[22] Gutmann, p. 214.

[23] PBS interview with Secretary West, 11 Sep 1997.

[24] Department of the Army, Human Relations Action Plan, *The Human Dimensions of Combat Readiness*, September 1997.

[25] Gilmore, "Sexual Harassment Reports Released," Army News Service, 5 Sep 1997.

issue and it will be addressed as such."[26] A set of parallel initiatives known as Character Development XXI—already under development before the events at Aberdeen—included revision of the Army's leadership manual FM 22-100, renamed Army Leadership, Be, Know, Do (31 August 1999), which defined Army values and introduced a new officer evaluation report to ensure that officers were rated according to how their behavior reflected Army values.[27] General Reimer also distributed throughout the Army a commercially produced video called "Living Army Values," intended for professional development, and a mandatory ethical climate assessment survey for senior leaders. The character development program included a "consideration for others" initiative, modeled after a similar program at the U.S. Military Academy at West Point designed by the Commandant of Cadets, Brig. Gen. Robert Foley, who had served on the Senior Review Panel. This initiative was designed to develop a culture in which people treated one another with dignity and respect. Finally, the CSA distributed a sexual harassment prevention chain-teaching program throughout the Army to define *sexual harassment*, identify the role of the chain of command and the reporting procedures available, reemphasize the Army's zero-tolerance approach, and ensure leader involvement in human relations training.[28]

Of special interest to those concerned for the future of mixed-gender BCT was the action plan's proposal for TRADOC to add an extra week to the eight-week basic training program to "include additional human relations training that inculcates Army values, appropriate behavior, and team-building."[29] On 27 August 1997, even before the public release of West's action plan, a process action team was formed at TRADOC headquarters to develop a concept plan for the revision of not only BCT but AIT and the One Station Unit Training program, as well. On 7 October 1997, the TRADOC staff presented recommendations to TRADOC commander Gen. William W. Hartzog that BCT be extended to nine weeks beginning on 1 October 1998. The hours of instruction would be increased by fifty-four hours and would include human relations, core values, and Army heritage and traditions. Each week of basic training would be assigned a value that would be integrated into every subject trained. (The values included loyalty, duty,

---

[26] Ibid.

[27] The Aug 1999 version of FM 22-100 superseded the 31 Jul 1990 edition and incorporated a number of other leadership doctrinal manuals.

[28] Human Relations Action Plan, pp. 4-1–4-3, 6-1, 12-1; Gilmore, "Chief Speaks on Sexual Harassment," Army News Service, Mar 1997.

[29] Human Relations Action Plan, pp. 8-1, 12-2; TRADOC Public Affairs Office, "BCT Expands One Week to Include Values-Based Training," 23 Apr 1998.

respect, selfless service, honor, integrity, and personal courage, allowing the use of the acronym LDRSHIP). In addition, seventy-two hours at the end of each cycle would be devoted to a warrior field training exercise, similar to the Marine Corps' Crucible, concluding with a rite-of-passage ceremony.[30] In part to demonstrate the importance the Army placed on training recruits, the position of deputy commanding general for initial entry training (at the level of lieutenant general) was created at TRADOC headquarters. On the following day, the same recommendations were approved by Army Chief of Staff Reimer.[31] In separate actions, the Army Recruiting Command began issuing sexual harassment policy cards, which provided all new recruits with the steps they should take if they experienced sexual harassment, and the Department of the Army began studying the feasibility of distributing a values card to basic trainees, setting out the Army's newly approved values. Eventually, all Army soldiers received values cards.

Congress Investigates

As Secretary West's review panel collected its data and the IG performed inspections at various installations, Congress began its investigation of the sexual harassment charges that increasingly appeared to pertain to the entire Army. The Senate Armed Services Committee held its first hearing on 4 February 1997. During the hearing, several senators called for the military (all services except the Marine Corps, which did not have gender-integrated training) to reevaluate mixed-gender training at the basic level to determine whether it should continue. Secretary of Defense Cohen informed lawmakers that while defense officials would consider whether training men and women together created an environment conducive to sexual abuse, he and most Army leaders did not think that mixed-gender training was the problem. Rather, the problem rested with failures in leadership. Secretary of the Army West advised senators that "when our women in the armed forces do their jobs, they do them in a gender-integrated atmosphere....It seems not only foolish but perhaps a waste of taxpayers' money to train them separately in a way different from the way they will be expected to perform."[32]

Army Chief of Staff Reimer told the committee, "[women] serve with men, why not introduce them at the very start?" He further stated:

---

[30] Training hours were also added to AIT. The drill sergeant program of instruction went through an extensive revision.

[31] Options briefing to CSA, 8 Oct 1997.

[32] Kozaryn, "Senators, DoD Leaders Address Sexual Harassment," Armed Forces Information Service, 6 Feb 1997.

"We would not have as good an Army as we have right now if we didn't have the females in. We need to attract from the broad base of society and take the best people to be a part of our Army."[33]

The undersecretary of Defense for Personnel and Readiness, Edwin Dorn, told the committee "unwanted sexual behavior in the military has declined in recent years according to a DoD survey [1995, see above]. The good news is that sexual harassment isn't as bad as it used to be. The bad news is that it's still a major challenge."[34]

The secretaries of the Navy and Air Force also testified and defended their mixed-gender basic training programs.

Following the hearing, two members of the committee, Sen. Charles Robb (D-Va.) and Sen. Olympia Snowe (R-Maine), expressed their views. Senator Snowe asked why, in the Navy and Air Force, where fewer women were excluded from specialties, there was less sexual discrimination than in the Army and Marine Corps, where more women were excluded. Could it be that more women were needed in the chain of command? Senator Robb added, "If we are not going to put them [females] in actual ground combat...it may make sense to take a look at whether or not we ought to train them in the same way." He continued, "There may be some legitimate reasons for not putting people, particularly in that very vulnerable stage, in exactly the same type of training at exactly the same time....If we are not going to make the ultimate decision to put women in the front lines in so-called 'hand-to-hand' combat, then maybe we ought to take a look at whether or not we ought to train them for those same types of activities."[35]

On the day after the Senate hearings, the action moved to the House of Representatives, where Del. Eleanor Holmes Norton (D-D.C.) of the Congressional Caucus on Women's Issues addressed the full House. It was her understanding that after the revelations at Aberdeen, the Army's senior leaders had "unequivocally pledged that they would never go back to discriminatory training of men and women." However, she expressed her understanding of the Senate testimony of Army Chief of Staff Reimer, who suggested that the Army might be open to a reexamination of gender-integrated training. Norton explained, "It is totally unacceptable to move back to the dark ages when there were two armies, one for men and one for women. The Army itself has field tested [mixed-]sex training and found that it improves the performance and morale of women with no negative effect on unit cohesion.... To

---

[33] Ibid.
[34] Ibid.
[35] PBS interview with Senator Robb, 4 Feb 1997.

the Army brass, I say, don't throw in the towel. Above all, don't throw the towel at women. They can die together; they can train together."[36]

Shortly thereafter, House Speaker Newt Gingrich (R-Ga.) directed the House National Security Committee, which had primary responsibility over the Department of Defense, to fully investigate the issue of sexual misconduct in the military services.[37] Committee chairman Rep. Floyd Spence (R-S.C.) and ranking minority party member Rep. Ron Dellums (D-Calif.) assigned the investigation to three members of the Subcommittee on Personnel: Rep. Steve Buyer (R-Ind.), Rep. Jane Harman (D-Calif.), and Rep. Tillie Fowler (R-Fla.). The House investigative team sought to determine whether the armed forces could police themselves or whether "extraordinary avenues" needed to be created to address allegations of sexual misconduct in basic training.

Before the team could begin that investigation, allegations were raised that coercive investigative practices by Army investigators had led to inappropriate pressure on individuals to make false allegations or to make admissions in violation of due process and Fifth Amendment rights against self-incrimination. The House investigators also took up the matter. The latter charges would eventually lead to action by the Army Criminal Investigation Command.[38] Meanwhile, the investigative team traveled to Aberdeen Proving Ground, the Navy's Great Lakes Training Center, Ill.; the Air Force's basic training center at Lackland Air Force Base, Tex.; Marine Corps basic training at Parris Island, S.C.; and Fort Leonard Wood, which was one of the Army's basic training centers directly involved with the allegations of sexual misconduct.[39] The team was heavily armed with information gleaned from the court-martial of one of the Aberdeen drill sergeants.[40]

Debate continued in the Senate. Sen. Richard J. Santorum (R-Pa.), a member of the Senate Armed Services Committee, was asked whether men and women should be trained separately: "I have suggested to military chiefs that they take a look at separate training of male and

---

[36] "Sexual Harassment and the Army," *Congressional Record*, 5 Feb 1997, H.R. 290.

[37] The House National Security Committee was, for a brief time, a new name for the House Armed Services Committee.

[38] "The National Security Committee's Investigation of Sexual Misconduct in the Military," *Congressional Record*, 13 Mar 1997, p. E-481. This investigation had not begun as late as Oct 1997.

[39] Women may receive basic training for the Marine Corps only at Parris Island, although there is also a training center near San Diego.

[40] House of Representatives, 105th Cong., 1st sess., Committee on National Security, statement of Reps. Steve Buyer, Jane Harman, and Tillie Fowler following the court-martial verdict of Staff Sgt Delmar G. Simpson, 29 Apr 1997.

female recruits....I don't think we know at this point whether female troops trained along with male troops are 'better trained' than female troops trained separately. However, some evidence has been cited that incidences of improper conduct toward female troops are less frequent when those troops train separately."[41]

Senator Santorum was also asked what he saw as the role of Congress in affecting the sexual harassment policies of the armed services and in writing legislation that would force the sexes to be trained differently. His response was one popular with members of both political parties, particularly Republicans: "It is possible for the Congress to set policy in training our military recruits; however, I am not sure that this is the wisest course. The Congress should set standards for performance and conduct but should be very careful not to micro-manage our armed forces."[42]

By mid-May, the debate in the Senate over sexual misconduct had polarized on the issues of national security, social equality, and military readiness, with sexually integrated training of recruits increasingly the focus. Sen. Thomas Slade Gorton III (R-Wash.) called for a "thorough review" of integrated training in the military and took advocates of mixed-gender training to task: "The situation needs to be examined with a dispassionate attitude, and it greatly complicates our task if well-meaning advocacy groups in our country make the assumption that anyone who calls for a thorough investigation of the viability of gender-integrated training and operational roles is, per se, a bigot, is against equal treatment and opportunity, and is trying to roll the clock back because of his or her narrow vision."[43]

Apparently not seeing the contradiction in his own arguments, which were anything but dispassionate, Gorton continued by pointing out that there was no body of evidence that a force trained on a gender-integrated basis performed better than all-male units.

Irrespective of evidence, the highly charged debates over mixed-gender training of recruits continued in the committees and subcommittees of both legislative bodies and on the floor of the House and Senate. On 1 October 1997, several days after the release of the IG and the Senior Review Panel reports to the secretary of the Army, the House Committee on National Security's Subcommittee on Military Personnel held a hearing on the reports. In attendance were the reports' authors (see above), three members of the oversight team appointed to follow

---

[41] PBS, "Senate Armed Services Forum," 6 Mar 1997.
[42] Ibid.
[43] Senate, "Sexual Conduct, Training, and American National Security," 20 May 1997.

the Army's investigation, General Reimer, and Secretary West. During the hearing, most discussion focused on TRADOC's scarce resources, a result of the drawdown of forces after the Cold War ended, and the effect that situation might have on remedying the Army's problems. Discussion also included problems with the equal opportunity system and the ratio of drill sergeants to recruits. A few minutes were devoted to the problems of mixed-gender training, although at that time, there seemed to be little or no sentiment for abandoning the program.[44] Meanwhile, the Pentagon and the Army tried to define what policies would best fit the needs of military training in the future.

Department of Defense Investigates, June–December 1997

As debate on Capitol Hill intensified and Congress and the Pentagon awaited the reports of Secretary West's review panel and the IG's report, Secretary of Defense Cohen announced, on 27 June 1997, the appointment of a Federal Advisory Committee on Gender-Integrated Training and Related Issues, better known for the name of its chairperson, Sen. Nancy Kassebaum Baker (R-Kans.). Cohen gave the eleven-member (five retired military, six civilian) "Kassebaum Baker panel," which would report directly to him, six months to look at the viability and desirability of gender-integrated training at both the BCT and AIT levels and to observe morale, discipline, and physical training standards.[45] The defense secretary was deeply concerned about bills that had recently been introduced in both houses of Congress to outlaw the training of men and women together at the basic level (discussed below). In fact, he and Secretary of the Army West, Army Chief of Staff Reimer, the chief of Naval Operations, and the Air Force chief of staff sent letters

---

[44] Department of the Army Reports on and Corrective Actions Related to Recent Cases of Sexual Misconduct and Related Matters, 105th Cong., 1st sess., Hearing before the Military Personnel Subcommittee of the Committee on National Security, House of Representatives, 1 Oct 1997. A bill had been introduced earlier in the House to eliminate gender-integrated training (GIT) but did not succeed (see below). TRADOC commander Gen William W. Hartzog was not in attendance, because he was out of the country at the time.

[45] A major problem for the Army leadership was the seeming inability to explain to the media and some members of Congress that AIT, which had been integrated for years (e.g., since 1976 for the Air Force), was not considered "basic training." The Kassebaum Baker panel was made up of private citizens and retired military officers and included Lt Gen Robert Forman, a former TRADOC deputy commander for Training; Condoleezza Rice, provost of Stanford University; Carolyn Ellis Stanton, former vice chairwoman of DACOWITS; and John Dancy, former broadcast journalist with NBC News.

addressed to members of the House National Security Committee voicing united support for mixed-gender training. Defense Secretary Cohen wrote Representative Harman (see above), "This force is the product of the way the services train. It would be a mistake to mandate limits that could have a negative effect on the cohesion...and readiness of our forces." West and Reimer declared, "Turning the clock back to gender-segregated training will result in unrealistic training, which degrades readiness." Each of the correspondents stressed that his respective service "trained the way we fight."[46]

After visiting seventeen military sites representative of all the services; engaging in discussions with hundreds of recruits, recruiters, instructors, first-term military members, and supervisors in operational units; and receiving dozens of briefings by senior-level service personnel on their training programs and policies, the Kassebaum Baker panel submitted its report to Secretary Cohen on 17 December 1997. In most respects, the report is startlingly like the report that resulted from the 1995 study by the U.S. Army Research Institute for the Behavioral and Social Sciences, which was not released until February 1997 (see Chapter III). This similarity suggests that the problems with basic training were present and recognized on some levels long before the allegations at Aberdeen. The Kassebaum Baker report recommended more and better training for recruiters and the cadre; more female instructors; decreased emphasis on monetary incentives in recruiting in favor of more motivational themes, such as patriotism; and tougher physical fitness requirements with consistent standards for male and female recruits. The report also recommended the elimination of "no touch, no talk" policies recently put in place by some commanders and tough punishments for false accusations regarding sexual harassment and misconduct.[47]

None of these recommendations was especially disquieting to the military leadership. However, two of the thirty recommendations, all adopted unanimously, alarmed and disappointed the senior leadership of all the services, especially the Army. The panel recommended that

---

[46] Armed Forces Information Service, "Defense Leaders Support Mixed-Gender Training," Jun 1997; Office of the Assistant Secretary of Defense, Public Affairs, DoD News Briefing with Secretary Cohen, 27 Jun 1997; Office of the Assistant Secretary of Defense, Public Affairs, news release, "Secretary Cohen Announces Task Force Members," 27 Jun 1997.

[47] "Report of the Federal Advisory Committee on Gender-Integrated Training and Related Issues to the Secretary of Defense," Executive Summary, 16 Dec 1997. According to some of the "no-talk, no-touch" policies, men were directed not to look at a woman for more than three seconds nor were recruits allowed to speak with members of the opposite sex.

male and female recruits and trainees in BCT and AIT be provided separate barracks rather than divided single barracks. The panel did not recommend separate campuses. Even more threatening to the Army's new mixed-gender training program was the recommendation to end gender-integrated training in all the services: "The committee recommends that the Army, Navy, and Air Force organize all their operational training units by gender in platoons [Army], divisions [Navy], and flights [Air Force]. The committee believes this will recapture the cohesion, discipline, and team-building of living and training together as an operational unit."[48]

Why did the panel make these recommendations knowing that all the affected services would likely reject them? Perhaps most important was the frequency with which members encountered in the field a concern for the quality of BCT graduates who were arriving for AIT. Recruits were said to be undisciplined and disrespectful, with little military bearing and poor technical skills. Likewise, drill sergeants at the gender-integrated Army basic training sites complained of an "inordinate amount of time spent investigating or disciplining male/female misconduct."[49]

Recruits of both sexes complained that they could not work as a team in barracks that contained members of more than one operational unit who tended to compete with each other. The committee concluded that "gender integration at the operational training unit level is causing confusion and a less cohesive environment." In addition, the "no talk, no touch" rules that brought the sexual harassment policy down to its most enforceable level and the buddy system had rendered training more concurrent than truly integrated. Trainers said that they did not know their recruits as well as they had when training units that were not mixed, and, thus, were together twenty-four hours a day.

The panel also concluded that separating recruits would provide a better environment for teaching military values and professionalism. Also persuasive were the "vast majority" of recruits, AIT trainees, and newly assigned service members who expressed their belief that gender-integrated training had "gone soft." Perhaps an influencing factor for the panel was the "observed impressive levels of confidence, team-building, and esprit de corps" they found in the all-female training platoons at the Parris Island Marine Corps base, an atmosphere that was not observed in the other services' training units.[50]

---

[48] Ibid., p. 13.
[49] Ibid., p. 11.
[50] "Report of the Federal Advisory Committee on Gender-Integrated Training and Related Issues to the Secretary of Defense," Executive Summary, 16 Dec 1997, pp. 12–14.

Upon receiving the report, Cohen, without acceptance or rejection, forwarded the Kassebaum Baker recommendations to all the services for their review and reaction, to be received by the Department of Defense in ninety days, or by 16 March 1998. Meanwhile, the Defense Advisory Committee on Women in the Services (DACOWITS) entered the fray to declare strongly that most men and women in the military believed that the sexes should be mixed more, not less, in basic training. In a report of its own, based on visits to twelve training schools at nine installations, the Pentagon advisory panel concluded that team-building and professional relationships across gender lines were "service objectives that could not be met through artificial barriers imposed on trainees of different genders." Both the Kassebaum Baker panel and DACOWITS came under fire.[51]

Three months later, in early March 1998, Defense Secretary Cohen received the comments of all the services and held a news briefing on 16 March to discuss the reports and his plans. The service chiefs agreed with approximately 95 percent of the Kassebaum Baker panel recommendations—the important exceptions being the recommendations for separate barracks and the abandonment of mixed-gender training at the basic level. Collectively, they recognized a number of deficiencies and agreed to make changes in those areas. Cohen made it clear that he would address only basic training, not advanced training, because "if we can fix the problem at basic, we believe it will have a continuum of influence throughout the advanced and military occupational specialty assignments."

Those changes included more female recruiters and trainers; better selection processes for trainers and better definition of their authority; the encouragement of professional relationships without gender-based policies, such as "no talk, no touch"; more emphasis on patriotism in recruitment advertising; a greater emphasis on core military values; and more consistent training standards between the genders. Cohen directed the service chiefs to report back within thirty days with detailed plans for implementation. He also directed additional action in the key areas of basic training: leadership and the value placed on training; the rigor of training; and recruit billeting.[52]

With regard to leadership, Cohen hoped to counter, by increased rewards and incentives, the widely held perception that a training assignment was detrimental to one's military career. On the matter of more

---

[51] "Pentagon Panel Finds Military Personnel Want More Mixed Training," *Boston Globe*, 21 Jan 1998.

[52] Office of the Assistant Secretary of Defense, Public Affairs, news briefing, 16 Mar 1998.

rigorous training, he wanted the physical fitness standards reevaluated and toughened. With regard to billeting, Cohen stopped short of accepting the Kassebaum Baker panel's recommendation, instead calling for "separate areas" for male and female trainees and better supervision. He denied that cost was the deciding factor in determining how recruits should be housed. The services were to report back to him in thirty days on the steps being taken to achieve these goals. Once the recognized deficiencies were corrected, Secretary of Defense Cohen would "evaluate the need to alter small-unit gender integration during basic training." Cohen would accept information from any or all of the services; the decision would be his alone.[53]

On the following day, the House Subcommittee on Military Personnel of the Committee on National Security held a hearing to receive the reports of each of the services and to hear statements from Elaine Donnelly, president and founder of the Center for Military Readiness; D. Michael Duggan, deputy director for National Security, Foreign Relations Division of the American Legion; and Brig. Gen. Evelyn Foote (Ret.), vice chairwoman of Secretary West's senior review panel of 1997 and an experienced commander of a gender-integrated basic training battalion from 1978 to 1982.

Representing the Army was Vice Chief of Staff Gen. William W. Crouch. Crouch informed the subcommittee that the Army had "already acted upon many of the recommendations on the basis of previously conducted internal reviews." Many of these "improvements" dealt with drill sergeants: new selection procedures, psychological screening, and increased human relations training. In addition, as discussed above, basic training would be extended to nine weeks to "better instill Army values in our new soldiers," and a field training exercise would be added. New physical fitness standards would bring requirements for men and women closer.

Despite Secretary Cohen's claim that the services had agreed to 95 percent of the Kassebaum Baker panel's recommendations, the Army only "partially concurred" with nine recommendations. Partial concurrence seems to have meant that the service believed that the recommendation was a good idea, but it could not be implemented. First among the recommendations that came under this heading was the suggestion that the Army decrease monetary incentives in recruitment advertising (college tuition assistance). General Crouch told the representatives that "educational benefits remain the top reason for enlistment, and we see value in advertising this powerful incentive." Further, the Army

---

[53] Ibid.

rejected the recommendation that the Delayed Entry Program be used as a mandatory program to enhance physical fitness in enlistees on the grounds that injuries would have legal implications. In addition, the Army agreed that recruiters with the lowest attrition among their enlistees should be rewarded, but no system for determining such awards was in place at the time.

On the panel's recommendations that the number of drill sergeants be increased, that their training be improved, and that the three drill sergeant schools be consolidated, Crouch told the subcommittee that TRADOC believed the ratio of drill sergeants to recruits to be correct. The command had also added thirty-eight hours of human relations training (values, training of mixed-gender units) to the drill sergeant program of instruction and was studying "several consolidation options" because of the potential costs of such an action. As for separate barracks for male and female recruits, the Army would meet the intent of the recommendation by providing "separate and secure housing," but the service believed separate barracks would negatively affect unit cohesion, teamwork, and discipline and would exceed the limitations of the Army's training base infrastructure.

On the recommendation that punishment for false accusations of sexual misconduct be severe, the Army would trust its commanders to judge such incidents and to act in accordance with the Uniform Code of Military Justice. The Kassebaum Baker panel had also recommended an increase in the number of female recruiters. The Army responded that this was a goal, but a competing demand for female drill sergeants made it necessary to carefully balance requirements for quality female NCOs. The Army strongly rejected the two remaining recommendations—the abandonment of mixed-gender training for combat support and combat service support soldiers and the shifting of more training to initial entry training to reduce the training requirements of operational units. The latter recommendation was a response to criticism that soldiers arrived at their first unit assignments without the proper training or skills, making it necessary for unit commanders to assume part of the training burden.[54]

The Navy and Air Force statements to the subcommittee were similar to those of the Army but addressed each service's unique mission. Both services defended mixed-gender training at the basic level. The Navy based its defense on the necessity to train recruits in an

---

[54] Statement by Vice Chief of Staff Gen William W. Crouch, USA, before the Subcommittee on Military Personnel, Committee on National Security, House of Representatives, 105th Cong., 2d sess., Army Response to the Kassebaum Baker panel report, 17 Mar 1998.

environment similar to the conditions aboard ships, where there is little or no personal privacy.[55] The Air Force spokesman affirmed the service's support for "training as we operate day to day.... We view the challenge of instilling discipline as a leadership issue, not an organizational issue."[56] The Marine Corps, with no gender-integrated training at the basic level, was much less defensive. "Gender-segregated training provides an environment free from latent or overt sexual pressures, thereby enabling new and vulnerable recruits the opportunity to focus on and absorb Marine [Corps] standards of behavior."[57]

Elaine Donnelly minced no words: "The entire issue has been reduced to a vacuous and illogical slogan: Gender-integrated training is necessary because we must train as we fight. No one seems to notice that if we fight as we train—burdened with unprecedented disciplinary problems, gender-normed double standards, high drop-out rates, and other problems that our potential enemies don't have—America's armed forces will be in deep trouble."[58]

Donnelly praised the Kassebaum Baker panel report for its independence and criticized recent studies that defended mixed-gender training, which she claimed, were done in an atmosphere "where dissent from administration policy is simply not an option." Donnelly expressed her belief that the Kassebaum Baker panel report did not go far enough, but it was important "in that it refutes this culture of political correctness in favor of military correctness. It is a small step on the path to political courage." Finally, she urged Congress to "find out why the Kassebaum Baker panel voted as they did."[59]

Another witness, D. Michael Duggan, told lawmakers that "decades of experience with separate gender training worked for the veterans of World War II, the Korean War, Vietnam War and Persian Gulf War." He

---

[55] Statement of Vice Chief of Naval Operations Donald L. Pilling, USN, before the Subcommittee on Military Personnel, Committee on National Security, House of Representatives, on Kassebaum Baker panel recommendations and DoD response, 17 Mar 1998.

[56] Statement of Vice Chief of Staff Gen Ralph E. Eberhart, USAF, before the Subcommittee on Military Personnel, Committee on National Security, House of Representatives, on Kassebaum Baker panel recommendations and DoD response, 17 Mar 1998.

[57] Statement of Assistant Commandant of the Marine Corps Gen Richard I. Neal before the Subcommittee on Military Personnel, Committee on National Security, House of Representatives, on Kassebaum Baker panel recommendations and DoD response, 17 Mar 1998.

[58] Testimony of Elaine Donnelly before the Subcommittee on Military Personnel, Committee on National Security, House of Representatives, on Kassebaum Baker panel recommendations, 17 Mar 1998, p. 1.

[59] Ibid., pp. 2, 14.

also urged Congress not to allow special interest groups to compromise the indoctrination of recruits. Like Elaine Donnelly, he used the Marine Corps basic training program as the model that the other services ought to follow.

Finally, the subcommittee heard Brig. Gen. Evelyn Foote (Ret.), past commander of a mixed-gender basic training battalion during the Army's first attempt at training men and women together in the late 1970s and early 1980s (see Chapter II). Foote had also served as vice chairwoman of Secretary West's Senior Review Panel shortly after the reports of sexual assault and rape in initial-entry training at Aberdeen Proving Ground. She told the subcommittee that while she agreed with many of the Kassebaum Baker panel recommendations, she could not support the abandonment of gender-integrated basic training, nor did she believe that separate barracks were essential. Her experience as commander of the 2d Basic Training Battalion, U.S. Army Training Brigade, Fort McClellan, Ga., had taught her that the soldiers she had to be most concerned about were not the recruits but the company commanders, drill sergeants, and instructors who did not want to change. Brigadier General Foote reminded the subcommittee that all soldiers take the same oath at enlistment and have the same duty to live up to the oath, thereby requiring the same training.

In April 1998, the services began to send the required responses to Secretary Cohen in accordance with his 16 March 1998 directive to address cadre leadership issues, recruit billeting, and training rigor. At that point, controversy about the training of men and women together in basic training had reached its peak. To avoid as much criticism as possible, Cohen retained G. Kim Wincup, former assistant secretary of the Army for Manpower and Reserve Affairs and former assistant secretary of the Air Force for Acquisition, to conduct an independent review of the services' responses to both the Kassebaum Baker panel report and the secretary's additional guidance.[60] By 1 May 1998, Wincup had completed his review. Cohen promptly released the services' response and Wincup's assessment for internal Pentagon use. On 8 June, the information was provided to the news media and the public.[61]

In general, the Army's response changed little from that of March 1998. The service planned to place less emphasis on college tuition assistance in its recruitment advertising campaigns and more on patriotic values. In addition, plans were to place one female recruiter in each

---

[60] Early in 2000, Wincup served as vice chairman of the Congressional Commission on Servicemembers and Veterans Transition Assistance.

[61] DoD news briefing, 8 Jun 1998.

large recruiting station. Further, there were now more details on ways to improve the quality of life and offer greater recognition for BCT trainers. To raise the status of trainers, drill sergeants would receive additional special duty pay and promotion points. In answer to criticism that the Army physical fitness test, as well as the standards applied to the confidence course, were "gender-normed" to favor women, the Army promised a more rigorous physical training point scale to be applied throughout the service. Point totals for men and women would continue to vary only to the extent of physiological (age and gender) differences between the sexes.[62] Finally, the service announced implementation of a seventy-two-hour continuous field training exercise culminating in a "rite of passage."[63]

Only regarding the issue of separate and secure recruit housing did the Army spell out its plans, likely hoping to save gender-integrated training, which was once again a priority for discussion in Congress. Independent sleeping areas and latrines would be required. Men and women would be separated by bays, floors, or fire-safe barrier walls. Separate entrances to the living area would be provided, and sleeping area entrances would be secured from outside access. All doors would have alarms. Supervision of the barracks would be controlled by the chain of command; NCOs would supervise the barracks twenty-four hours a day; and the charge-of-quarters would be required to be a drill sergeant during sleeping hours. In addition, there would be periodic checks of the living areas by duty officers and NCOs at the company, battalion, and brigade levels. Soldiers would also be required to sleep in a physical fitness uniform.[64]

If the above conditions could not be met, separate structures would be required. In this latest report to the Department of Defense, none of the services except the Marine Corps, which segregated the sexes, mentioned the possibility of eliminating mixed-gender training. Such a suggestion was probably not an option. In any case, it is clear that the proposed solutions to the alleged incidents of sexual misconduct had by this time become focused on the single issue of gender-integrated

---

[62] On 1 Oct 1998, the sit-up requirements became the same for men and women; the push-up and run requirements were toughened but continued to vary according to age and gender. Standards are determined by the U.S. Army Physical Fitness School at Fort Benning, Ga.

[63] Undersecretary of Defense, "Report on the Responses of the Armed Services to the Federal Advisory Committee on Gender-Integrated Training and Related Issues and Additional Direction by the Secretary of Defense," 1 May 1998, passim.

[64] Ibid., p. 8; deputy chief of staff for Training, TRADOC. In the spring of 1999, surveillance cameras were installed in mixed-gender barracks.

training. It is equally clear that Secretary of Defense Cohen strongly supported the policy that each service should be allowed to decide what was best for its members.[65]

Perhaps a partial explanation of Secretary Cohen's stalwart rejection of the Kassebaum Baker panel report with regard to mixed-gender basic training rests in a General Accounting Office (GAO) report of 16 March 1998—published the same day that Cohen released the military services' reaction to the Kassebaum Baker panel recommendations. The GAO report was a response to a request from democratic Representative Dellums, ranking minority member of the House National Security Committee,[66] that the GAO investigate the methodologies of the senior review panel, the Kassebaum Baker panel, and the DACOWITS' report. Dellums wondered how the three committees have come to such disparate conclusions. Perhaps their methodologies were flawed.

The GAO found that Secretary West's review panel provided "ample support for making conclusions and recommendations." The Kassebaum Baker panel, on the other hand, had failed to "systematically collect the same information from all groups" and had not documented the information generated or explained how it reached specific conclusions and recommendations. As for DACOWITS, its representatives focused on broad issues and on determining what issues the group would concentrate on in the future. They had reached no conclusions or made any recommendations on basic training based on the data collected. Thus, the GAO report concluded with regard to both the Kassebaum Baker panel and DACOWITS' reports that "problems with the methodology limit the usefulness of the reports."[67]

As officials of the Army and the other services considered and implemented their responses to the Defense Department, an effort was ongoing in the House of Representatives to legislate gender-integrated training out of existence. On 8 May 1997, Rep. Roscoe G. Bartlett (R-Md.) introduced a bill to amend Title 10 of the U.S. Code "to require that recruit training in the Army, Navy, Air Force, and Marine Corps be conducted separately for male and female recruits." The bill also would allow, beginning 31 December 1997, only male officers to command and NCOs to serve as drill instructors in male training units; the same segregation would apply to female units. The bill, which had 121 cosponsors, was referred to the House National Security Committee,

---

[65] Associated Press, "Military to Divide Sexes in Barracks," 9 Jun 1998.

[66] Democrats lost control of the House of Representatives in the off-year elections of 1994.

[67] GAO, "Gender Issues: Analysis of Methodologies in Reports to the Secretaries of Defense and the Army," Washington, D.C., 16 Mar 1998.

of which Representative Bartlett was a member, for executive comment from the Department of Defense. On 4 June 1997, the Defense
Department returned an "unfavorable" response. The following day,
Bartlett withdrew his bill, perhaps in anticipation of the appointment of
the Kassebaum Baker panel, which was announced on 27 June 1997.[68]
Bartlett, however, calling Army basic training "social engineering [that]
has proven a failure" and accused Cohen of "a lack of intestinal forti-
tude" and of abdicating his responsibilities by letting the services decide
training policy.[69]

Meanwhile, on 16 May 1997, republican Sen. Olympia Snowe
introduced a bill in the Senate, presumably to head off any success that
Bartlett might have in the House: "The secretary of Defense shall take
such action as is necessary to ensure that the armed forces continue gender integration of the training programs of the armed forces."[70]

Senator Snowe had been highly critical of the Army for the behavior she believed it had allowed to occur at Aberdeen, but she had been
and would continue to be a strong advocate of mixed-gender training.

There the matter stood until the following spring. Then, during the
2d session of the 105th Congress, Representative Bartlett reintroduced
his bill in the House as an amendment to the defense authorization bill.
This time, the proposed legislation was cosponsored by Representative
Buyer, the chairman of the Military Personnel Subcommittee of the
National Security Committee, and Rep. Gene Taylor (D-Miss.), a member of the committee. It may be remembered that Buyer was chairman
of the investigative team that House Speaker Newt Gingrich appointed
shortly after the charges made by female trainees at Aberdeen and other
locations. Buyer and Taylor were joined by Rep. Tillie Fowler and Rep.
Jane Harman in an unsuccessful attempt to strike the Bartlett amendment from the authorization bill. The House approved the measure with
a voice vote on 20 May 1998.

Approximately one month later, the action moved once more to the
Senate. On 24 June 1998, Sen. Sam Brownback (R-Kans.) offered an
amendment to the fiscal year 1999 defense authorization bill that would
have ended mixed barracks beginning on 15 April 1999, with completion scheduled for 2001. On the following day, Sen. Robert C. Byrd
(D-W.Va.) offered an even stronger amendment that would incorporate

---

[68] House of Representatives, "Military Recruit Training Restoration Act of 1997,"
105th Cong., 1st sess., *Congressional Record*, 8 May 1997, H.R. 1559; Kozaryn, "Defense
Leaders Support Mixed-Gender Training," Armed Forces Information Service.

[69] *Express News*, Mar 1998; *New York Times*, 17 Mar 1998.

[70] Senate, "To Ensure the Continuation of Gender-Integrated Training in the Armed
Forces," 105th Cong., 1st sess., *Congressional Record*, 16 May 1997, S. 760.

the language of the Brownback bill and incorporate the House measure to end mixed-gender training in small units. The military leadership and women's rights groups were suddenly alarmed. The service chiefs (except the Marine Corps) all wrote letters to the office of the chairman of the Senate Armed Services Committee, Sen. Strom Thurman (R-S.C.). Chief of Staff of the Army General Reimer wrote, "we should begin Day One of a soldier's life in a gender-integrated training environment." Former Army Chief of Staff and president of the Association of the United States Army Gen. Gordon R. Sullivan wrote, "Young men and women, entering the Army from an environment where the genders will be routinely mixed...do not benefit from a brief period of artificial separation."[71]

Secretary Cohen gave the issue a pragmatic twist: "If Congress is going to mandate separate barracks, then I hope that they will also appropriate the dollars to fund it."[72] Senior female officers in all the services descended on senior staffers in protest. In an atmosphere reminiscent of the women's rights groups' intense lobbying in 1991 concerning the repeal of the combat exclusion law, current and former members of DACOWITS and women's rights advocates crowded Senate offices and corridors. At the same time, the American Legion and the Veterans of Foreign Wars publicly announced their support for the Brownback amendment.

At the same time and, perhaps, ironically, organizers were planning for top defense officials, lawmakers, and female officers to assemble at the Women in Military Service for America Memorial at Arlington Cemetery to commemorate the golden anniversary of President Harry S. Truman's signing of the Women's Armed Services Integration Act on 12 June 1948, giving women a permanent role in the military services. Air Force Brig. Gen. Wilma L. Vaught (Ret.), president of the Women's Memorial Foundation, perhaps said it best as senators from both sides of the aisle debated military personnel policy: "Here we are fifty years later fighting the same battles."[73]

This time, leaders of the Army, Air Force, and Navy prevailed as both the Brownback and Byrd amendments failed by fairly wide margins.[74] In his closing remarks, Senator Byrd claimed to be left with

---

[71] Senate, "Brownback and Byrd Amendment No. 2978," 105th Cong., 2d sess., *Congressional Record*, 24 Jun 1998; "Byrd Amendment, No. 3011," 25 Jun 1998; Minerva Bulletin Board, summer 1998.

[72] Tom Brown, "Senate Set to Consider Segregating Barracks," *Baltimore Sun*, 10 Jun 1998.

[73] Ibid.

[74] Center for Military Readiness, notes, Jul 1998.

"one looming fact…our military is not an equal employment opportunity commission." The three services would continue, for the present, to train men and women together at entry level and recruits would all live in the same structures with more built-in safeguards. But the services' relief was to be short-lived. Failing to pass the successful House measure in the Senate, conservatives inserted language in the pending fiscal year 1998 defense authorization bill to accept the findings of the Kassebaum Baker panel, which supported gender-separate basic training. Supporters of gender-integrated training persuaded the Senate Armed Services Committee to delay any action until the results of yet another panel, the Congressional Commission on Military Training and Gender-Related Issues, had completed its investigation. As it had in 1992 with the Presidential Commission on the Assignment of Women in the Armed Forces (see Chapter III), assignment of a commission to provide recommendations let both the military services and the Congress off the hook, at least temporarily. The new Commission on Military Training and Gender-Related Issues, which Congress established shortly before the release of the Kassebaum Baker panel report of December 1997, is known as the Blair Commission, named after chairwoman Anita K. Blair, and referred to as the gender-integrated, or GIT Commission.[75]

Blair Commission

From the beginning, the Blair Commission was contentious and controversial. Although established in November 1997, it was not organized to conduct business until well into the spring of 1998. The major problem at this stage was congressional infighting over the appointment of the commission's ten members. Five commissioners were to be appointed jointly by the chairman and ranking minority party member of the House National Security Committee (Rep. Floyd Spence (R-S.C.) and Rep. Ike Skelton (D-Mo.). The other five were to be appointed by the chairman and ranking member of the Senate Armed Services Committee, Senator Thurman and Sen. Carl Levin (D-Mich.). Thus, the commission started out in an atmosphere of divisive partisan politics that would worsen over time. Those on each side of the basic training gender question did not want to accept the choices of the other side.[76]

---

[75] Statement and Status Report of the Congressional Commission on Military Training and Gender-Related Issues, 17 Mar 1999. Hereafter cited as GIT Statement and Status Report. The commission was established under Title V, Subtitle F, of the National Defense Authorization Act of FY 1998, PL 105-85, 18 Nov 1997.

[76] *Washington Times*, 19 May 1998.

The list of the members of the Blair Commission makes fascinating reading, because it was the makeup of the membership that made the body arguably the most divisive in the history of Congress. Indeed, there was dissension from the beginning over whether there ought to be only ten members or eleven—as the Kassebaum Baker panel had and as the Senate Armed Services Committee had recommended.[77] Elected chairwoman of the commission was Anita K. Blair, a lawyer and executive vice president of the Independent Women's Forum.[78] Blair, one of the Senate's choices, was then serving on the Board of Visitors of the Virginia Military Institute and was actively involved in the attempt to bar acceptance of women to the state-supported military college.[79] Also the choice of the Senate was the commission's vice chairman, Air Force Col. Frederick F.Y. Pang (Ret.), who had served as assistant secretary of Defense for Force Management and as assistant secretary of the Navy for Manpower and Reserve Affairs. Colonel Pang had close ties to the Pentagon and to the administration of President Bill Clinton.

As a third member, the Senate Armed Services Committee chose Maj. Gen. William Keys (Ret.) of the Marine Corps, who was a veteran of the Vietnam and Persian Gulf wars. The two remaining Senate choices were academic professionals. Charles Moskos, Ph.D., a professor of sociology at Northwestern University, had previously served on several committees that looked at the issue of women in the military. He was also the author of many books and articles on the American military establishment and the military services' "don't ask, don't tell" policy on gays and lesbians. Last, there was Nancy Cantor, Ph.D., provost of the University of Michigan, whose membership appears to have been a compromise between Senators Levin and Thurman.[80]

The members chosen by the House of Representatives National Security Committee were also a mixed group. Lt. Gen. George R. Christmas (Ret.) was a former deputy chief of staff for Manpower and Reserve Affairs of the U.S. Marine Corps. Representing the Army was

---

[77] TRADOC, Congressional Significant Activities, 26 Jun 1997.

[78] The Independent Women's Forum was an organization known for its antifeminist agenda.

[79] Anita Blair, subsequent to chairing the GIT Commission, served on the American Conservative Union's 2000 Committee for a conservative platform. Beginning in Aug 2001, she served in President George W. Bush's administration as deputy assistant secretary of the Navy for Personnel Programs. She also was executive director of the panel to investigate the sex scandals at the Air Force Academy in 2003–2004.

[80] House of Representatives, "GIT Statement and Status Report," 106th Cong., 1st sess., 17 Mar 1999. Declaration by professor Charles Moskos, U.S. District Court for the District of Columbia. Dr. Cantor is, at this writing, chancellor of the University of Illinois at Urbana-Champaign.

the retired command sergeant major of Forces Command and former drill sergeant Robert A. Dare Jr. Also chosen was Thomas Moore, director of International Studies at the Heritage Foundation and a graduate of the Citadel military college in South Carolina; and Barbara S. Pope, who served as assistant secretary of the Navy during President George H. W. Bush's administration. The House also chose Mady W. Segal, Ph.D., professor of sociology at the University of Maryland. Professor Segal was frequently a consultant to the armed services, most recently to Secretary West's Senior Review Panel on Sexual Harassment (see above). In 1996, President Clinton appointed Segal to the Board of Visitors of the U.S. Military Academy.

This, then, was the group of private citizens who received a charter from Congress to once again study gender-integrated training. The charter (PL 105-85) tasked the commission as follows:

- Review requirements and restrictions regarding cross-gender relationships of members of the armed forces.
- Review the basic training programs of the Army, Navy, Air Force, and Marine Corps.
- Make recommendations on improvements to those programs, requirements, and restrictions.[81]

The lengthy and specific charter asked many of the same questions that had guided the Kassebaum Baker panel, but it was clearly written to favor a conclusion that gender-integrated training should be eliminated in basic training for all the services. For example, the Army, Navy, and Air Force based their defense of integrated training on a "train as you will fight" or "train as you will operate" rationale. The commission was asked to "Assess whether the concept of 'training as you will fight' is a valid rationale for gender-integrated basic training or whether the training requirements and objectives for basic training are sufficiently different from those of operational units so that such concept, when balanced against other factors relating to basic training, might not be a sufficient rationale for gender-integrated basic training."[82]

The charter's focus could be interpreted as finding proof that there was no advantage to mixed-gender basic training rather than requiring comparison of the two modes of training enlistees. And, once again, the focus was almost entirely limited to training of new recruits, despite the

---

[81] Charter, Congressional Commission on Military Training and Gender-Related Issues (Blair Commission), FY 1998, Defense Authorization Act, spring 1998.
[82] Ibid.

fact that allegations of sexual misconduct came primarily from soldiers in AIT.[83]

Even in its formative stages, the commission created controversy. On 19 May 1998, at the commission's second closed-door organizational meeting, four members (Pope, Segal, Pang, and Cantor) abruptly walked out after failing to stop some of the chairman's staff appointments. Anita Blair's opposition to mixed-gender training was well known on Capitol Hill. All four dissenting members announced their intentions to resign, presumably to win staff appointments more to their liking. When the action threatened to eliminate the commission entirely, democrats convinced the four to remain, because the existence of the commission was their rationale for delaying a vote to enact the Kassebaum Baker panel findings, which would have eliminated integrated training.[84] In any case, the incident underscored the deep ideological divide within the group.

Originally, the commission was directed to submit an initial report to the appropriate committees on 15 April 1998, with a final report due in October 1998. Because of the commission's rocky start, the due date for the initial report was changed in the fiscal year 1999 Defense Authorization Act to 15 October 1998, with the final report to be submitted on 15 March 1999. Those dates, too, were allowed to slip. On 17 March 1999, the commission was finally able to submit a status report to the House National Security Committee, with a final report to be filed in mid-April 1999. However, it was July 1999 before the report was finally finished.[85] It was clear from the beginning, as one member of the commission observed, that not a single commissioner had entered the investigation with an open mind. Gender-integrated basic training had become a single-issue phenomenon about which no one was neutral. Even more remarkable, perhaps, was the rare cooperation by the Army, Navy, and Air Force, as the three military services fought to save their mixed-gender training programs, if not always for the same reasons.

The reasons for the delays were procedural, as well as ideological. By June 1998, the commission was ready to begin work. During the summer of 1998, liaison officers from each of the services arranged the commissioners' inspection tours at seventeen initial-entry training

---

[83] The Blair Commission also studied the military services' policies concerning fraternization and adultery, but these issues are beyond the scope of this study.

[84] Rowan Scarborough, *Washington Times*, 19 May 1998. The charter gave the chairwoman sole authority to appoint an executive director and three additional staff members.

[85] Charter, Congressional Commission on Military Training and Gender-Related Issues (Blair Commission).

sites representing all the military services. Also in June, the commission received detailed readiness briefings from each of the military services. During the fall and winter months of 1998–1999, the well-funded commission also spoke with hundreds of witnesses, conducted several surveys, and sent interrogatories with quick turnaround dates to the Department of Defense, the services, and major commands.[86] For example, on 21 December 1998, the commission sent to TRADOC headquarters a set of nineteen questions regarding basic training, each of which required days of research and the preparation of numerous tables and graphs, with a requested due date of 7 January 1999.[87]

The Blair Commission also conducted twelve days of hearings, from October 1998 until the end of January 1999.[88] In support of the hearings, Chairwoman Blair sent interrogatories to all the military services that included such questions as the following:

- Assess the feasibility and implications of conducting basic training…at the company level and below through separate units for male and female recruits, including the costs and other resource commitments.
- Assess the feasibility and implications of requiring drill instructors for basic training units to be of the same sex as the recruits in those units.

In their replies, all the services addressed prohibitive costs and a negative effect on unit cohesion if men and women were trained separately in basic training. For example, Louis Caldera, secretary of the Army in 1998, estimated that to house gender-separate platoons would initially cost $217 million. He expressed his concern that gender-separate training would replace teamwork with competition and take away important combat arms experience from female recruits, who would no longer be trained by male drill sergeants. Besides, separate training of female units instructed by female drill sergeants would require 245 additional female drill sergeants, which the Army could not provide without taking women from operational units. The Air Force argued that if sexual conduct could not be controlled during basic training in a tightly restricted environment, control would be even more difficult to attain in operational units.[89]

---

[86] There were a total of ten surveys and research projects.

[87] E-mail msg, Deputy Chief of Staff for Personnel to TRADOC DCST, 21 Dec 1998.

[88] Reports of time spent with hearings varies. The Mar 1999 presentation to the House Armed Services Committee refers to twenty-one days of formal hearings.

[89] Secretary of the Air Force F. Whitten Peters to Chairwoman Blair, Jan 1999. Ninety-nine percent of USAF career fields were open to women.

The Blair Commission's initial findings of 17 March 1999 resembled the earlier Kassebaum Baker panel report, but the commission's recommendations were very different.[90] Whereas the Kassebaum Baker panel had voted unanimously to recommend elimination of gender-integrated training, the Blair Commission voted six to one with three abstentions to allow the services to continue training men and women together under current programs. The Marine Corps could continue gender segregation in basic training. It was clear that during the many months the commission had studied the issue, no minds had been changed. Commissioner Moore of the Heritage Foundation voted no to the recommendation to continue mixed-gender training; commissioners Blair, Moskos, and Keys abstained; the remainder voted with the military services. Before the final report, issued in July 1999, Blair and Keys changed their votes to no. Professor Moskos continued to abstain. The final recommendation was short and to the point:

> The commission concludes that the services are providing the soldiers, sailors, airmen, and marines required by the operating forces to carry out their assigned missions; therefore, each service should be allowed to continue to conduct basic training in accordance with its current policies. This includes the manner in which basic trainees are housed and organized into units. This conclusion does not imply the absence of challenges and issues associated with the dynamics found in a gender-integrated basic training environment. Therefore, improvements to initial entry training that have been made by the services, or are currently being considered, must be sustained and continually reviewed.[91]

The majority explained its position: "It is leadership and command climate that determines the success of initial entry training. The degree of separation has less of an impact on the outcomes of basic training than does the behavior of the leaders…. It will continue to be these leaders and their command environment that sustain the mission readiness of the services."

The six commissioners who voted in favor of gender-integrated training explained that when asked about their major concerns, leaders did not mention gender unless specifically asked. Instead, the

---

[90] The account in the text of the commission's report is based on the report to the House Armed Services Committee of 17 Mar 1999, which does not differ from the final report in the essential details. The 31 Jul 1999 report contains, however, 2,700 pages and is available only in compact disk form. The final report is in four volumes, containing Findings, Recommendations, Transcripts, Legal Consultants' Reports, Research Projects, and Studies.

[91] Report to the House Armed Services Committee, 17 Mar 1999, p. 27.

commissioners' concerns revolved around sustainability based on personnel shortages and increased operational tempo (OPTEMPO). When asked what the commission should tell Congress, one officer declared "personnel or OPTEMPO, fix one or the other."[92] Col. Frederick Pang, vice chairman and one of the commissioners in the majority, chided the members of the minority, who called for experimentation before continuing mixed-gender training: "They want to kick the can down the road by putting on the backs of the military further studies and experiments—just to keep an agenda alive."[93]

Despite defeat, opponents of mixed-gender training did not give up easily. Rep. Roscoe Bartlett, the original author of the successful House bill to end gender-integrated training, vowed to continue to fight. On the same day that the commission appeared before the House Armed Services Committee, Sen. Sam Brownback once again rose in the Senate to ask "if gender-integrated training is necessary, then why do we not hear reports of difficulties arising from the segregated training of the Marines?" He requested that the Senate Armed Services Committee hold formal hearings upon submission of the commission's final report, to once again consider segregated barracks.[94] Chairwoman Blair declared that "the tension between quality and equality is apparent everywhere."[95] Commissioner Thomas Moore, noting that the debate still had not been laid to rest and that the commission had failed to provide a true test, exonerated himself: "If our forces should fail the test, or if our soldiers, sailors, and airmen suffer needless deaths because the White House, Pentagon, and Congress felt it more important to carry out an unprecedented social and cultural revolution, then I hope all present will remember the opportunity we had to remedy a fatal error, remember what was said and done here today, and remember who was responsible. My conscience, at least, will be clear."[96]

In sum, the Army's mixed-gender basic training program in the 1990s had been in existence only two years when the events at Aberdeen Proving Ground and other Army installations threatened its future. Army leaders and, indeed, those of the Navy and Air Force had reason to hope the controversy would not continue after the Blair Commission

---

[92] Ibid., pp. 28–29. Opponents claimed that the word had gone out throughout the Army, at all levels, that gender issues were not to be discussed.

[93] Report to the House Armed Services Committee, 17 Mar 1999, statement of Commissioner Fred Pang, p. 2.

[94] *Congressional Record*, 17 Mar 1999, p. S. 2838.

[95] Report to the House Armed Services Committee, 17 Mar 1999, statement of Chairwoman Anita Blair, p. 3.

[96] Ibid., statement of Commissioner Thomas Moore, p. 4.

report. It was not to be. As the gender-integrated training programs continued, so did the criticism. As Commissioner Moore had observed, the polarization of opinion over the issue meant that it would not easily be put to rest.[97]

---

[97] When the documents of the Blair Commission were deposited in the National Archives, they contained a large number of disciplinary reports collected during visits to basic training sites. No analysis of the reports was included in the final report to Congress because of an internal dispute over what conclusions to draw. A lawyer for the commission and an opponent of integrated training claimed, "the reports show that basic training has become more of a summer camp than preparation for war. . . only 11 percent of drill instructors and other trainee supervisors made positive comments about the state of recruit training." Scarborough, *Washington Times*, 18 Jan 2000.

V

Into the Twenty-First Century
• •

A New Political Era, 2001

The results of the Congressional Commission on Military Training and Gender-Related Issues (Blair Commission) were not widely known or discussed. The gender-pure advocates had no reason to celebrate their failure to end mixed-gender basic training. The successful champions of gender-integrated basic training at the Pentagon had no desire to call attention to the matter—the less publicity, the better. For fifteen months after publication of the final report in late July 1999 and with the Clinton administration still in the White House and the Pentagon, the debate over training men and women in the same small units dwindled. An occasional amendment to a defense authorization bill routinely went down in defeat. The issue received little media attention, with the exception of several articles authored by Elaine Donnelly, president and founder of the Center for Military Readiness, and Anita Blair, executive vice president of the Independent Women's Forum.[1] Then, in the 2000 presidential election, republican George W. Bush defeated his democratic opponent, and the gender-pure advocates, now calling themselves "pro-military," were rejuvenated.

Beginning in the summer of 2001, a new wave of attacks on the Department of Defense and its advocacy of gender-integrated training began. As might have been expected, opponents based their arguments on the Kassebaum Baker panel report of 1997 rather than on the Blair Commission report of 1999. The Kassebaum Baker panel report recommended eliminating mixed-gender training in basic combat training

---

[1] Independent Women's Forum interview with Anita Blair the day after the final commission report was submitted, Article No. 484, 1 Aug 1999. Elaine Donnelly, "Women in Combat: Time for a Review," *American Legion Magazine*, Jul 2000, pp. 12–15. Donnelly accused the services of "gender-norming [adjustment]" in integrated training and asked whether they would accept the concept on the battlefield or at the Army-Navy football game.

(BCT), while the Blair Commission report recommended no changes. As more studies appeared to demonstrate that there was little difference in the results of gender-pure and gender-integrated training, opponents of integrated basic training began to ask: If it makes no difference, then why have integrated training? Sometimes it was suggested that cost might be a reason, although the Army did not have definite cost comparisons between the two modes of training. In any case, as soon as the Bush administration had its own team in place, the attacks began, encouraged by what some perceived as President Bush's campaign promise to return to single-sex training for new recruits.[2]

The Heritage Foundation, a conservative think-tank based in Washington, D.C., led the way via the print media in mid-July. Noting that the president's 2002 budget contained a request for funds to "arrest the decline in military readiness," policy analyst Jack Spencer expressed his belief that the problem demanded more than additional funding: "Washington must rethink recent social policies that are affecting the readiness of U.S. armed forces to the detriment of national security. The first step should be to end gender-integrated officer and enlisted basic training, which numerous studies show is resulting in lower standards, increased misconduct, and declining morale."[3]

Returning to a well-worn argument, Spencer maintained that male trainers often applied less stringent physical training requirements to female trainees for fear of charges of sexual harassment. The Marine Corps, with its gender-separate basic training and its claims of trouble-free recruiting and retention, was held up as a model for the other services. Spencer called on Secretary of Defense Donald H. Rumsfeld to adopt the conclusions of the Kassebaum Baker panel and order—effective immediately—an end to mixed-gender basic training.[4]

In late July 2001, during a panel discussion sponsored by the Heritage Foundation, Rep. Roscoe G. Bartlett (R-Md.), who first submitted a bill in the House of Representatives to end gender-integrated basic training (which later died in the Senate) and who vowed to keep trying, explained his position. Why, he asked, did no women play professional sports? "It is simply because God made men and women different." Bartlett also believed women created a national security risk in that a woman taken prisoner was more susceptible to rape and abuse than were male soldiers: "I don't want women serving in any capacity

---

[2] Candidate Bush only characterized the issue as "something to look into."

[3] Jack Spencer, "Why the Social Experiment of Gender-Integrated Training has Failed," Heritage Foundation Executive Memo No. 758, 18 Jul 2001.

[4] The author could find no evidence that Secretary Rumsfeld's office made any reply.

where they could become prisoners of war (POW), because they will be raped and tortured. When this happens in front of our men, our men will break. Lastly, I don't think that the hand that rocks the cradle should be shooting the heads of the enemy."[5]

In early August 2001, fourteen self-styled pro-defense groups, joined by the Veterans of Foreign Wars, signed a letter addressed to Defense Secretary Rumsfeld pressing him to follow through on what they believed was President Bush's campaign pledge to reexamine the practice of training male and female recruits together in the same units and urging him to reverse this policy: "In our view, military policies should encourage discipline, not sexual misconduct. There is ample evidence that training men and women together complicates and detracts from the training mission. Our members hope that you will act quickly to end this and other demoralizing personnel policies that have vitiated discipline and morale."[6,7]

When the *Washington Times* asked the secretary what his actions would be, Rumsfeld replied that no one "within the military" had raised military social issues with him. Besides, he replied, he was far too busy with the policy study known as the Quadrennial Defense Review, which would retool the armed forces to confront twenty-first-century threats. Elaine Donnelly, a former advisor to President George H. W. Bush on defense matters, voiced her disappointment: "My concern is that without proper attention, the agenda set in motion during the Clinton administration is moving ahead like a battleship on autopilot and something has to be done to change course."[8]

Hoping to gain more information, the *Washington Times* interviewed President George W. Bush's national security adviser, Condoleezza Rice, who had served on the Kassebaum Baker panel and had informed reporters during the presidential campaign that Bush would look at separating men and women during BCT. Rice replied that Rumsfeld had too few people to study policy, but she added, "I do think it is an issue that will come back." Conservatives voiced their hope that when the service chiefs learned that the president favored a switch in policy, they would be agreeable to separating the sexes.[9]

---

[5] Jason Pierce, "Coed Basic Training Hurts Military, Experts Say," Cybercast News Service, 25 Jul 2001.

[6] The letter quoted candidate Bush as telling *American Legion Magazine* (Jan 2000), "experts tell me that we ought to have separate basic training facilities."

[7] Signers included leaders of the Center for Security Policy, the Coalitions for America, the Center for Military Readiness, and Concerned Women for America, among others.

[8] Rowan Scarborough, "Pentagon Urged to Separate Sexes: Groups Want Bush Promise Upheld," *Washington Times*, 7 Aug 2001, pp. A-1, A-16, quotation, p. A-16.

[9] Ibid.

A return to separate training for men and women in combat support and combat service support military occupational specialties (MOSs) clearly was not a priority for Rumsfeld. High on his agenda was the strengthening of civilian influence in the Department of Defense, which he perceived to have eroded during the Clinton administration. As a result, relations between the secretary and some senior military leaders in the Pentagon were strained. In mid-2003, Defense Secretary Rumsfeld asked for the resignation of Secretary of the Army Brig. Gen. Thomas E. White (Ret.), in part, because he believed White sided more with the military chiefs than with the civilian leadership. If Rumsfeld had wanted to challenge the service chiefs, gender-integrated training offered a fertile field of battle. Nevertheless, he did not seem to be attracted to that sort of confrontation as the first Bush term concluded.

In any case, opponents of mixed-gender training did not abandon their efforts. In late August 2001, *Insight* magazine, a sister publication of the *Washington Times*, published a series of articles on the subject. An article by Representative Bartlett set forth the opponents' case. Calling gender-integrated basic training "deficient in all relevant respects," Bartlett saw the program as the result of "the political agenda of the Clinton administration to reward its feminist supporters." He saw dead soldiers, grieving families, and a weakened military as the "ultimate price that our nation pays when its leaders make political correctness a priority over national security." Congress, he maintained, had "a long history of "granting great deference to the military concerning the details of training." He continued: "It's past time to end our country's second experiment in gender-integrated basic training of our military recruits. Must we wait before its failure shows up in more unnecessary sexual scandals or worse, body bags in our next conflict? The cost is just too high."[10]

Capt. Lory Manning USN (Ret.), director of the Women in the Military Project at the Washington, D.C.–based Women's Research and Education Institute, wrote a rebuttal to Bartlett's essay. With regard to the Kassebaum Baker panel report, on which Bartlett's conclusions were based, Manning explained that the recommendation for a return to gender-pure basic training had nothing to do with toughening training or preventing sexual encounters. The panel's rationale had been that recruits who were together twenty-four hours a day (that is, those who shared sleeping quarters) developed better teamwork than those who did not. Given that men and women in mixed-gender units

---

[10] Rep. Roscoe G. Bartlett, "Has the Military's Gender-Integrated Basic Training Been Successful? No," *Insight*, 20 Aug 2004.

never shared sleeping quarters, the committee recommended gender-segregated training to allow for shared sleeping quarters with gender-pure units. As for the major problems that a number of previous studies revealed, especially in the Army, Captain Manning offered the following explanation:

> Beset by downsizing, reduced budgets and changing missions in the aftermath of the Cold War, [the services] were not paying attention to basic training. They were not monitoring the caliber of basic-training staffs or the training given to those staffs. In an attempt to save personnel and funding, they had...understaffed key training facilities that stretched drill sergeants too thinly and had failed to provide enough oversight by experienced senior enlisted personnel and officers. Segregating the genders would correct none of these failures.

Not really hoping to change any minds among opponents of integrated training, Manning nonetheless observed that most of the crew involved in the attack on the U.S.S. *Cole* (DDG-67), who performed so magnificently, were products of gender-integrated basic training.[11]

The opponents of mixed-gender training saw an opportunity in the tragedy of the attacks on the World Trade Center, the Pentagon, and Pennsylvania on 11 September 2001. A group, meeting at the Center for Military Readiness concluded that the attacks "could have been prevented if the last administration had not shrunk the budget and size of the military or given in to social demands from women's groups....The state of the military is rotten."

Elaine Donnelly added, "One good thing to come from the 11 September attacks, however, may be the military's return to basics."[12]

Meanwhile, Representative Bartlett was living up to his vow to continue efforts to eliminate gender-integrated training for Army recruits. On 6 September 2001, Bartlett and the vice chairman of the House Armed Services Committee, Duncan Hunter (R-Calif.), sent a letter to Training and Doctrine Command (TRADOC) commander Gen. John N. Abrams via Secretary of the Army Thomas White.[13] Bartlett and Hunter informed Abrams of their desire to restructure basic training, that is, to return to gender-pure basic training, as a part of Army transformation

---

[11] Capt Lory Manning USN (Ret), "Has the Military's Gender-Integrated Basic Training Been Successful? Yes," *Insight*, 20 Aug 2004.

[12] Kelly Beaucar Vlahos, "CIA Recreating Military Might," Fox News, 19 Oct 2001.

[13] Ltr, Bartlett and Hunter to White, 6 Sep 2001; Ltr, Joe G. Taylor Jr., TRADOC OCLL, to Abrams enclosing Bartlett and Hunter's ltr to White, 24 Sep 2001; Abrams's handwritten forward to deputy chief of staff for Operations and Training.

efforts begun in the fall of 1999.[14] Asserting that TRADOC's basic training program was out of line with Defense Secretary Donald Rumsfeld's stated philosophy in matters of force management, Bartlett and Hunter bluntly stated their views of gender-integrated training: "There is abundant evidence that the Army's gender-integrated basic training program for enlisted personnel is an expensive social program that has created problems, while generating minimal confirmed results. We hope you agree that the process of transformation must include a fresh look at this, as well as other personnel policies that vitiate morale, discipline, and overall readiness."[15]

The congressmen attached their plan, which they termed "Army Basic Training Initiative," to once again segregate the sexes in basic training.

In defense of their program, Bartlett and Hunter chose supporting phrases from both the Kassebaum Baker panel report of 1997 and the Blair Commission report of 1999 (see Chapter IV). TRADOC's existing program, they said, required constant sensitivity training, a situation that made remedial instruction necessary to compensate for lessons not learned. The only advantages TRADOC could point to were "associated with social and political objectives, rather than military value." They called on Abrams to "take immediate action to restore sound priorities." Gender-integrated training had not, they wrote, improved retention or recruitment, unlike with the Marine Corps and its single-sex basic training. Most of the arguments the congressmen advanced were the same that had been made dozens of times. The letter and its lengthy attachments were forwarded from the Pentagon to General Abrams on 24 September 2001.

Abrams, in turn, forwarded the documents to his deputy chief of staff for Operations and Training, Maj. Gen. Raymond D. Barrett Jr., who had just arrived at TRADOC headquarters from Fort Jackson, S.C., where he was commander.[16] Abrams also asked that Barrett's successor at Fort Jackson, Maj. Gen. David Barno, be consulted. Maj. Gen. Raymond Barrett's staff prepared a briefing for the congressmen using as a model a briefing prepared at Fort Jackson for Secretary White (see below). Generally, the briefing expressed support for the results of the

---

[14] "Army transformation was an ambitious and multifaceted program begun by Army Chief of Staff Gen Eric Shinseki (1999–2003) to bring the Army into the twenty-first century."

[15] Bartlett and Hunter ltr to Secretary of the Army Thomas White.

[16] Maj Gen Raymond Barrett signed into TRADOC on 15 Sep 2001 but did not start work until early Oct. His predecessor, Maj Gen John Brown, departed on 1 Sep 2001. Msg, Nelson C. Dodd to the author, 7 Oct 2004.

studies by U.S. Army Research Institute for the Behavioral and Social Sciences (ARI) and highlighted the history of gender-integrated training, compared the three services that supported gender-integrated training, spelled out the standards of the program, and noted feedback from the field. The conclusion was, as so often before, that mixed-gender BCT improved soldierization for female soldiers, while soldierization for male soldiers remained stable.[17]

There matters stood until late in 2002, primarily because of preoccupation with the dozens of issues raised by the 11 September 2001 terrorist attacks and the arrival at TRADOC of a new commander.[18] In addition, TRADOC's senior leaders believed Bartlett's views were not widely held in Congress and that "neither house wants the issue to be added to their plate...given all else that is going on."[19] Nevertheless, the Operations and Training staff was prepared to brief Representative Bartlett by late November. No record has come to light that the briefing ever took place, perhaps because of deteriorating relationships between Secretary White and the White House and Pentagon.[20]

TRADOC Gender-Integrated Training Assessment

Meanwhile, in December 2001, Secretary of the Army Brig. Gen. Thomas White directed TRADOC to conduct a gender-integrated training assessment to provide him with information to answer the continued criticism of the gender-integrated program. Secretary White asked unit leaders, "If you had a blank sheet of paper, how would you design initial-entry training (IET) today?" He claimed to have no predetermined outcome in mind.[21] To answer the secretary, TRADOC commander Abrams set up an IET Task Force at Fort Jackson, the largest of the Army's mixed-gender basic training sites. For background, the group consulted the ARI studies of 1993–1995, the Kassebaum Baker panel report, and the Blair Commission report—all discussed in Chapter IV. As similar efforts in the past had done, the task force conducted site visits, distributed surveys, and conducted interviews with officers, noncommissioned officers (NCOs), drill sergeants, and graduates. Unlike their predecessors, the members of this task force considered alternative gender-integrated training models. For example, they set up test models

---

[17] Point paper, Lt Col Edwin J. Kuster Jr. to General Abrams, 2 Aug 2001.
[18] Gen Kevin Byrnes took command of TRADOC on 7 Nov 2002.
[19] Maj Gen Raymond Barrett to Lt Gen Dennis Cavin, 1 Nov 2002.
[20] Secretary White resigned on 25 Apr 2003, when serious questions arose about his former relationship with Enron Corporation.
[21] White was a retired brigadier general appointed by President Bush in May 2001.

featuring integration at different levels—company, platoon, and squad. One model tested "phased integration" of squads. The task force concluded that all the models undermined teamwork and hurt the performance of female soldiers. As to phased integration, "failure to integrate on day one reinforces a perception that female soldiers have to *earn* their way onto the team." Further, basic training with gender-segregated units encouraged "negative attitudes that are reinforced in the name of competition between genders rather than against a standard."[22]

Surveys from fifteen installations (TRADOC and Forces Command) revealed overwhelming support for training men and women together in BCT at the squad level, to confirm the Army's "train as we fight" philosophy. It was that ideology that opponents of integrated training had rejected as naïve and damaging to readiness and national security. The task force admitted, however, that challenges remained. Women continued to sustain a disproportionate number of injuries. There was still a severe shortage of female drill sergeants. Perhaps most important was the public perception of a double standard in physical training requirements because women were required to do fewer push-ups and were allowed more time to complete the run requirement. (The latter was true of all the services, even the Marine Corps, with regard to pull-ups and the run). One slide in the presentation given to Secretary White on 22 March gave critics of gender-integrated training some ammunition. The task force slide specifically concluded that the Army's mixed-gender basic training was effective but not efficient, presumably because of the cost of billeting and the treatment of injuries.[23]

At the Army's defense, once more, of its gender-integrated training program, representatives Bartlett and Hunter—the same congressmen who had come close to demanding that TRADOC commander Abrams direct a return to gender-segregated basic training—teamed with Elaine Donnelly to make yet another attempt to kill the program. The two lawmakers requested that Donnelly compile a comprehensive summary of all the findings and recommendations of the various panels, commissions, studies, and so forth that dealt with gender-integrated BCT.

---

[22] Briefing to Secretary of the Army, U.S. Army Training Center at Fort Jackson, "Gender-Integrated Training," 22 Mar 2002, pp. 1–7.

[23] Ibid., p. 11. Meanwhile, from Dec 2001 through Apr 2002, a congressionally mandated survey showed that sexual harassment, coercion, and assault, as well as sexist behavior and unwanted attention, had declined in all the services since 1995, sometimes significantly. In general, the Army showed the most improvement; the Marine Corps, the least. The study was not released until February 2004. Rachel N. Lipari and Anita R. Lancaster, "Armed Forces 2002 Sexual Harassment Survey," Defense Manpower Data Center, Feb 2004.

Donnelly produced a heavily footnoted eighteen-page compendium of facts on the subject, including "dozens of compelling reasons why coed training should be ended without further delay" and making much of the Army's admission to "inefficiency" in gender-integrated training. The survey did not point out that most of the studies had generally supported integrated training.

Representative Bartlett placed Donnelly's summary in the *Congressional Record* on 11 June 2003. Subsequently, the Center for Military Readiness placed on its Web site a petition titled "Americans for the Military Petition to President George W. Bush," calling for the president to end integrated training immediately. The petition laid out all the previous arguments and added a plea that Bush not allow activists to use the capture, injury, and death of female enlisted soldiers in a support unit ambushed in Iraq in 2003 (Pfc. Jessica Lynch and others, see below) to argue for the inclusion of women in special operations, submarines, and land combat units. The petition also stated that the Army's TRADOC found integrated training to be "of no objective military value." The command never made such a statement.[24] The Web site carries no evidence that Bush received or read the petition.

DACOWITS and a New Era

The Defense Advisory Committee on Women in the Services (DACOWITS) had for years recommended that the Army sexually integrate crews for the multiple launch rocket system. The Army always refused. The panel had begun in the 1990s to recommend that women be allowed in special operations forces (SOF) rotary wing aviation units but had been unsuccessful. In 2000, DACOWITS added women's service on Navy submarines to the list of recommendations. The committee continued to strongly support mixed-gender training throughout the services, a program that they had been instrumental in establishing in Army BCT. Despite a lack of success, these actions were not likely to endear the committee to those who opposed all such policies. In fact, they had begun to see DACOWITS as a major stumbling block in rolling back feminist programs. During the Clinton administration, the advisory committee known, by its opponents, as the "Pentagon feminists," enjoyed support (or at least neglect) throughout the defense establishment.[25] With the advent of the Bush administration, the situation rapidly

---

[24] Center for Military Readiness, *Army Gender-Integrated Basic Training: Summary of Relevant Findings and Recommendations: 1993–2002*, May 2003.

[25] Center for Military Readiness, "Will Agenda of Clinton-era Pentagon Feminists Still Prevail?" 30 Oct 2002.

changed. Led once again by Elaine Donnelly and Rep. Roscoe Bartlett and with editorial support from the *Washington Times*, opponents of gender-integrated basic training (and of women on submarines) vowed to get rid of the fifty-year-old DACOWITS committee.[26] The situation became tense in June 2001 as the Bush administration continued to take shape. The Army submitted to DACOWITS a draft update, which consolidated Army regulations on women in direct ground combat and was intended for use by the Army in the field. Representative Bartlett became suspicious that the Army was "angling to allow female soldiers to serve closer to combat." As a result, on 28 June, he asked twenty-seven members of the House Armed Services Committee (twenty-four republicans and three democrats) to sign a letter addressed to Secretary of Defense Donald Rumsfeld, asking him to explain the Army's actions and "to conduct a thorough review of Army policies concerning the assignment of female soldiers." The Army insisted on several occasions that it was "not seeking any change to the 1994 [Department of Defense] policy" in regard to the ground combat rule: "This is simply a periodic update to Army regulations to reflect this standing policy." The Pentagon also defended the rule. Elaine Donnelly remained suspicious, insisting that DACOWITS was trying to manipulate the Army. Scarborough wrote: "Since 1994, DACOWITS has continued to press the services to open more roles to women. It also has asked for briefings to explain how the direct ground combat rule is used to designate a unit all-male."

The issue died down as the House Armed Services Committee turned its attention to other matters. But the battle lines were drawn, especially for Donnelly.[27]

In August 2001, another voice joined the opponents of DACOWITS and of gender-integrated basic training. Writing in a publication of the Heritage Foundation, policy analyst Jack Spencer was critical of what he saw as troops who had fallen "into the hands of people more interested in social experimentation than national security." He believed that "military readiness has been sacrificed on the altar of political correctness." He pointed to DACOWITS as the problem: "Most of the problem can be traced to the Defense Advisory Committee on Women in the Services.... Over the years, DACOWITS has morphed into a hotbed of feminism driven by the flawed theory that, were it not for artificial barriers to women, they would be interchangeable with men in all military tasks."

---

[26] Elaine Donnelly was a former member of DACOWITS who seldom voted with the majority.

[27] Quotations are from Rowan Scarborough, "Panel Queries Army's Plans for Women: Fears Change in Combat Rules," *Washington Times*, 5 Jul 2001, pp. A-1, A-12.

Spencer blamed what he called "the dumbing down of basic training" on DACOWITS: "[Forty-seven] percent of females in the military bail out before the end of their third year of service, compared to 28 percent for men. Perhaps that's because female soldiers resent how DACOWITS, in its zeal to make the military more female friendly, advocates policies that lead to lower standards."[28]

In October 2001, Defense Secretary Rumsfeld ordered a review of some thirty advisory committees that reported to Pentagon leaders. DACOWITS was one of the targeted organizations. At that time, officials conducting the review denied that DACOWITS had been singled out, as some DACOWITS supporters had charged. However, when the advisory committee had not received Defense Department cooperation for a long-scheduled executive committee meeting or for its regular spring conference, nor had a new chairman been appointed, the members and advocates voiced serious concerns. Army Brig. Gen. Evelyn Foote (Ret.), former Women's Army Corps (WAC) commander and vice chairman of the senior review panel after the alleged sexual assault at Aberdeen Proving Ground, Md., and president of the Alliance for National Defense, wrote in a memo to her organization: "[W]e consider DoD's behavior in this case cavalier at best, unprofessional at worst." In a separate letter to Secretary Rumsfeld, General Foote wrote, "I am deeply concerned and disturbed" that the future of DACOWITS is in doubt. Those responsible for personnel policy at the Pentagon generally declined to comment.[29]

A month later, women's groups attending a conservative political action conference called on Bush policymakers to put an end to the Pentagon's "politically correct social engineering projects," claiming that "a gender-integrated military drives up costs, complicates missions, and endangers lives.... Coed basic training must be brought to an end." At a press conference, Elaine Donnelly characterized DACOWITS as "a tax-funded feminist power base within the Department of Defense." A board member of the Independent Women's Forum also attacked DACOWITS: "When the nation watches the Super Bowl on Sunday, there will be no women on either team, for the obvious reason that men are stronger than women. And yet, we now send women into combat. A truth that we intuitively grasp and automatically accept in the sports arena, we blithely ignore and rationalize away for the military."[30]

---

[28] Spencer, "Heritage Foundation Views 2001," 9 Aug 2001.

[29] Vince Crawley, "Women's Group Concerned About Its Future," *Air Force Times*, 24 Dec 2001.

[30] Lawrence Morahan, "Women's Groups Blast 'Politically Correct' Pentagon Policies," Cybercast News Service, 31 Jan 2002.

In previous years, the DACOWITS charter always received an automatic two-year extension. However, in 2002, the renewal date of 28 February came and went without an extension. In addition, no new appointments had been made (all thirty-five members were Clinton appointees). Deputy Defense Secretary Paul Wolfowitz continued to study the situation in an effort to decide whether to side with the conservatives and let the organization die, to maintain the status quo, or to redefine the role of DACOWITS. Elaine Donnelly reminded her readers that if President Bush continued the Clinton military agenda, it "would be a huge disappointment of those military voters who made all the difference in the election." She continued, "With a serious war going on, Secretary Rumsfeld should not retain an extreme feminist committee pushing a radical agenda that has nowhere to go but over the edge. If the committee's charter is allowed to lapse, a sigh of relief will be heard at all military bases and on all the ships at sea."[31]

Regardless of conservative opposition to an advisory committee that strongly supported mixed-gender training, as well as combat assignments for women, DACOWITS had its supporters. On 27 February 2002, Wolfowitz met with Rep. Heather A. Wilson (R-N.M.), an Air Force Academy graduate and a retired Air Force officer, who served on the DACOWITS committee during the first Bush administration. The pro-defense representative warned Wolfowitz that if he did not renew the DACOWITS charter or if he scaled it back, then she would put up a "stiff fight." She also believed that military women needed a voice that was outside the chain of command.[32]

On 1 March, the Pentagon announced that it would not renew the DACOWITS charter; instead, a new charter would redefine the focus of the committee. Several days later (on 6 March), the reconstituted charter was released to the public. The Department of Defense official announcement termed it "improved with a broader focus." That was not the way DACOWITS supporters saw it. Twenty-two Clinton appointees were not reappointed, and a support staff of more than thirty service representatives was cut back to six.[33] Instead of two large conferences annually, the committee held two "business meetings" each year. The budget was cut by some $150,000. Perhaps most important, DACOWITS would no longer choose its subjects or its

---

[31] Scarborough, "Panel on Military Women in Peril," *Washington Times*, 28 Feb 2002, pp. A-1, A-14, quotation, p. A-14.

[32] Ibid., pp. A-1, A-14.

[33] Under the original charter, DACOWITS members served three-year terms, and approximately one-third were appointed annually, an arrangement that made the group somewhat apolitical.

itinerary—that would be done by the Pentagon, specifically by the Office of the Undersecretary of Defense for Personnel and Readiness. Most issues for discussion concerned family quality-of-life matters and their effects on recruiting and retention instead of a focus on the assignment of military women. The Pentagon could adjourn any meeting not deemed in the public interest. Members had to have military experience or be past or present military family members.[34] Membership was capped at thirty-five, but at the end of October 2002, only thirteen members had been chosen. Representative Wilson (see above), the only female veteran serving in Congress, denounced the new charter, especially the limits on membership to people with military experience and the Pentagon's increased control. She expressed her belief that DACOWITS would cease to be an advisory group that could "tell them things they don't want to hear."[35] She called the action "unfair to our servicewomen and unwise for the Department of Defense."[36]

DACOWITS' self-styled "pro-military" opponents were disappointed that the fifty-year-old organization had not been eliminated, but most were happy to take the new guidelines for the time being. That is, until Deputy Defense Secretary Wolfowitz and David Chu, undersecretary of Defense for Personnel and Readiness, chose a retired Marine Corps lieutenant general to head DACOWITS. Lt. Gen. Carol Mutter was on the council of advisors of Brig. Gen. Evelyn Foote's (Ret.) Alliance for National Defense, an organization that some conservatives considered feminist. Elaine Donnelly chided Wolfowitz and Chu for the appointment, asserting that they had not considered "the implications of her close association with the feminist Alliance for National Defense, an organization led by some of the most doctrinaire advocates of Clinton-era social engineering in the military. [This] raises questions about the direction the new committee will take under her leadership."[37]

The picture opponents painted of DACOWITS was similar to the image they promoted of gender-integrated training. Rep. Roscoe Bartlett accused the advisory committee of trying to "undermine the military's

---

[34] Scarborough, "U.S. Moves Women Away From Combat: Defense Panel to Shift From Combat Assignments to Readiness Issues," *Washington Times*, 6 Mar 2002, pp. A-1, A-8.

[35] Rick Maze, "Congress' Sole Woman Veteran Has Problems with DACOWITS' Plan," *Army Times*, 6 Mar 2002.

[36] Vlahos, "Pentagon Revamps Women's Military Panel," Fox News, 7 Mar 2002.

[37] Center for Military Readiness, "New DACOWITS' Chair Named," 30 Oct 2002.

effectiveness" through a focus on gender issues to the exclusion of all else.[38] Another opponent went even further:

> The DACOWITS recommendations from the past 10 years read like an act from *The Vagina Monologues:* sexual harassment directives as a constant refrain; lobbying for increased child care services; and, most critically, a persistent drumbeat for expanded combat roles for women. A recommendation from 1991 chastised the Marine Corps for continuing to use the slogan: "A Few Good Men." The previous year featured a suggestion that the secretary of the Air Force develop a maternity coat as a uniform option. Suggested new recruiting slogan: "A gynecologist on every aircraft carrier!"[39]

These are just two examples of the criticism DACOWITS received from opponents as they characterized the organization as a tool for radical feminists. Seldom was it noted that the group had only an advisory role. A careful reading of recommendations made by DACOWITS does not reveal a radical organization. For example, the reference to gynecologists on carriers was sent to the secretary of Defense, as follows: "DACOWITS recommends to the services that military women have comprehensive gynecological exams to include screening tests, as recommended by the American College of Obstetrics and Gynecology guidelines."

The language is not shrill but reflects a professional tone that one might expect from a group composed in large measure of academicians. There is little evidence that the advisory panel attempted to change the culture of the military, but neither the organization nor its advice was the same as it had been when DACOWITS was established in 1951. For instance, as late as February 2002, one critic took the advisory group to task over its success in having women's hygiene products made available at base stores because women, too embarrassed to ask, were having them mailed from home.[40] Supporters denied that the work of DACOWITS had been driven by a radical ideology. As one Defense Department official put it: "These are not a bunch of bra burners."[41] Another supporter remarked that the day of the foxhole was over. Whatever the image, the debate over DACOWITS revealed the same

---

[38] *Human Events*, 11 Feb 2002.

[39] Charmaine Yoest and Jack Yoest, "Booby Traps at the Pentagon," *Women's Quarterly*, winter 2002. Charmaine Yoest is project director of the Family, Gender, and Tenure Project at the University of Virginia.

[40] Scarborough, "Panel on Military Women in Peril," *Washington Times*, 28 Feb 2002, p. A-14.

[41] Crawley, "Women's Group Concerned About Its Future," *Air Force Times*, 24 Dec 2001.

polarization that governed any discussion about women in the military, be it gender-integrated training or women in combat. The hysteria that surrounded the terrorist attacks in September 2001 only made it worse.

The RSTA Debate

In the fall of 1999, Army Chief of Staff Gen. Eric Shinseki announced a major effort, termed "transformation," to guide the Army into the next century. A part of that effort was the new Reconnaissance, Surveillance, and Target Acquisition (RSTA) squadrons designed to be faster and more flexible than the existing ground reconnaissance units. The new squadrons were designed to be a part of six new Interim/Stryker Brigade Combat Teams to be developed for the twenty-first-century Army. The first RSTA squadron was activated on 14 September 2000, with the second scheduled to follow on 16 May 2002. The first unit, training at Fort Knox, Ky., included eight female soldiers. Immediately, Elaine Donnelly, Representative Bartlett, and members of Concerned Women for America, began to press Bush appointees at the Pentagon to make RSTA male-only. The Army's response was that there were no plans to change the units' mixed-gender status. The Clinton administration had not viewed the new squadrons as direct-combat units, thus women could be assigned to them. The new units were considered a very desirable MOS for new enlistees.[42]

The opponents of what they believed to be a "politically correct" decision and fresh from a victory over DACOWITS would not yield. They obtained a copy of a memorandum of August 2001 written by Maj. Gen. B. B. Bell, then commander of the Army Armor Center at Fort Knox, to the Pentagon, which appeared to prove their contention that the RSTA units would be engaged in direct ground combat, making the assignment of women to them illegal: "The RSTA squadron, in its entirety, is designed for full-spectrum operations with sustained contact with enemy forces. RSTA soldiers are equipped with a full range of weaponry, from individual to antitank missiles, and are prepared to engage threat forces to retain the commander's freedom of maneuver."[43]

Proponents argued that assignment to such a unit was "the kind of leadership and advancement opportunity long denied women." Opponents saw it differently: "This is no longer a power game where

---

[42] Scarborough, "Women Taken Out of Army Squads," *Washington Times*, 30 May 2002.

[43] Concerned Women for America, "Women Removed From Army RSTA Squadrons but Many Servicewomen Remain in Combat Positions," 5 Jun 2002.

ambitious women can try to advance their careers. This is a matter of life and death. Any claim that women are equal to men in combat settings is utterly irrational."[44]

Donnelly tenaciously lobbied Deputy Secretary of Defense Paul Wolfowitz and Undersecretary of Defense for Personnel and Readiness David Chu through the early months of 2002 on the grounds that the inclusion of women in the reconnaissance units was a move by Clinton policymakers to maneuver around the land combat ban, a move not reported to Congress, as required by law. She reminded the policymakers that the squadrons likely would conduct ground sweeps in such locations as Afghanistan. Finally on 26 April 2002, the Department of the Army directed TRADOC to stop assigning women to RSTA units and to realign those already assigned to less dangerous positions. The service's explanation was that the decision was made at the request of TRADOC because of "evolutionary changes in the mission and the operating environment." By the end of May there were no more women in the new squadrons.[45]

Why was this incident, which affected only eight female soldiers, important to the basic training program? The reversal on the assignment of women in the new units, coming directly on the heels of the changes in DACOWITS, created uneasiness among supporters of women in the military and mixed-gender basic training. One supporter, writing for the National Women's Law Center, said: "We're beginning to see a trend that I hope doesn't go any further than this." A supporter and member of the Alliance for National Defense remarked, "by itself it's a drop in the bucket, but it's sending a message that women aren't going to be able to do the jobs they were previously able to do." A survey of the literature reveals that those who wrote about the RSTA incident, proponents as well as opponents, linked the changes in DACOWITS to the reversal of gender integration in RSTA; many expressed either hope for elimination or concern for the future for the Army, Navy, and Air Force mixed-gender basic training programs during the administration of President George W. Bush.[46]

U.S. Military Women in Iraq

The presence of large numbers of women among the troops that captured Baghdad and toppled Saddam Hussein's government in the

---

[44] David Yeagley, "Women Warriors," *Front Page Magazine*, 17 Jun 2002.
[45] Scarborough, "Women Taken Out of Army Squads," *Washington Times*, 30 May 2002.
[46] Bret Ladine, "Army Unit to Bar Women," *Boston Globe*, Jun 2002.

spring of 2003 brought to the fore once more many of the issues that the military services—especially the Army—had struggled with to a greater or lesser degree since the inception of the all-volunteer force. The questions of the assignment of women and of women in combat, draft registration, sexual abuse of deployed female soldiers, physical readiness training, and the abuse of Iraqi prisoners all contributed to the debate about gender-integrated basic training and the issue of whether women should serve in the military at all. In addition, a new kind of warfare dictated major changes in the BCT program of instruction (POI).

The Army forces that deployed to Iraq in 2003–2004 were demographically different from those of the Persian Gulf War of 1991. There were twice as many Hispanic soldiers and slightly fewer African Americans. The average age was 27, a year older than in 1991. There were fewer high school graduates, down from 98 percent to 91 percent. Women made up 16 percent of active-duty Army personnel, as compared to 11 percent twelve years earlier.[47] By the summer of 2004, 15,500 servicewomen had deployed to Iraq.

Women would also be much closer to the action, partly as a result of the changes of 1994 (see Chapter IV) that allowed women to serve anywhere except in units below brigade level that were likely to engage in direct ground combat. According to the Army, at the beginning of the Iraqi invasion, 91 percent of MOSs were open to women. Since 1994, 260,000 new positions had opened to women. How close to the front a soldier would be placed depended on whether a certain occupation was on the banned (direct combat) list. However, the concept of "front lines" had largely disintegrated in Iraq by 2003, blurring the designation of forward positions and rear boundaries. The range of weapons had significantly increased, allowing target acquisition from much greater distances. It was strategically advantageous for the enemy to take out supply lines and communications centers where, coincidentally, women were more concentrated. And the threat was now multidimensional and could be launched from anywhere. In Iraq, there was no clear line in the sand.[48]

How women served in Iraq was a subject of intense scrutiny, especially among those critics who opposed women in the military or feared women might see combat. Many female soldiers served in traditional

---

[47] Tom Infield, "Today's Army Differs from Gulf War Force," *Philadelphia Inquirer*, 2 Feb 2003.

[48] Ibid.; Erin Q. Winograd, "Army Opening Most Air Defense Artillery Slots to Female Soldiers," *Inside the Army*, 3 Mar 2003; Stephen J. Blank, *Rethinking Asymmetric Threats* (Carlisle Barracks, Pa.: Strategic Studies Institute, Sep 2003), p. 6; Ann Scott Tyson, "The Expanding Role of GI Jane," *Christian Science Monitor*, 3 Apr 2003.

roles, but many more than in 1991 chose nontraditional MOSs. Women served with chemical and engineering companies, as ammunition carriers, and as aviators flying bombers, fighters, and helicopter gunships. Many women served with Military Police units, especially at security checkpoints. In early February 2003, TRADOC approved the opening of most field positions in Air Defense Artillery to women. Specifically, women would now be allowed to serve in Short Range Air Defense units and as Avenger battery commanders.[49]

Although the situation had changed rapidly, women were still barred from Infantry, Armor, Multiple Launch Rocket System (MLRS) batteries in Field Artillery, and SOF, such as Rangers, Green Berets, and Navy Seals. The commando forces would not allow women to fly their helicopters, although female pilots flew the same aircraft in the conventional branches. It was the question of women serving in MLRS batteries and piloting SOF helicopters (as well as serving on submarines) that brought DACOWITS into confrontation with the Defense Department and subsequently led to the organization being reformed. DACOWITS argued, "There is insufficient evidence that SOF rotary wing aviation crews collocate [operate together] with units involved in direct ground combat."[50] The commander of the U.S. Special Operations Command insisted, "...Direct action has always been a primary mission of SOF, and...involves direct ground contact."[51]

Perhaps the most severe criticism of the interrelated concepts of women in the military, women in combat, and mixed-gender basic training came as a result of events surrounding Pfc. Jessica Lynch. Lynch and other members of her 507th Maintenance Company from Fort Bliss, Tex., were attacked outside An Nasiriyah, Iraq, on 23 March 2003. One woman was killed, while Lynch and another female soldier became POWs. A much exaggerated rescue made Lynch front-page news for days. In general, the public reaction was the same as that accorded the women's male comrades. A majority appeared to believe as one writer did, that despite ground combat rules, "warfare rarely follows the textbook when it comes to when and where violence will erupt."[52]

---

[49] Winograd, "Army Opening Most Air Defense Artillery Slots to Female Soldiers," *Inside the Army*, 3 Mar 2003. Short Range Air Defense slots (532 of 10,000 assigned to Air Defense Artillery) in light divisions and those heavy divisions and corps positions in direct support of divisional maneuver elements remained closed to women.

[50] DACOWITS Recommendation, spring 2000.

[51] Scarborough, "Female Warriors Kept Off Ground for Special-Operations Missions," *Washington Times*, 24 Oct 2001.

[52] Martha Ackmann, "Restricting Women's Military Role Hurts All," *Newsday*, 8 Apr 2003.

However, among the opponents of mixed-gender training and women in combat, efforts resurfaced to change U.S. military policy on the assignment of women. Immediately after the story of the 507th became available through the media, Elaine Donnelly renewed the argument for a reversal of the Clinton administration's liberal approach to the assignment of MOS on grounds that it was essentially "unequal." Donnelly believed that the prospect of sexual abuse and rape gave women an unequal opportunity to survive. She contended that the Army had lowered physical training standards to ensure that women completed BCT and questioned the training in basic combat skills and weapons training for members of the 507th Maintenance Company. She also characterized an erroneous *Washington Post* article on Lynch's rescue as "a pro-women in combat media campaign."[53] President Bush chose not to join in the fray over women soldiers operating in combat zones. "I will take guidance from the United States military. Our commanders will make those decisions." The Senate and House Armed Services committees passed 2004 defense authorization bills without any amendments debated or passed to expand or restrict women's war missions. Despite the commander in chief's reluctance to get involved, the debate over how women should be trained and how they should be employed remained polarized. Those who opposed combat exclusion for women saw the Pfc. Jessica Lynch episode as proof of how women could perform in combat situations. To those who opposed the assignment of women to infantry and submarine military specialties, Lynch was a "victim of the PC [politically correct] military career myth sold to young women through feminist propaganda."[54]

The arrest of seven enlisted Military Police guards for abuse of detainees at Abu Ghraib prison in Iraq provided still more opportunities for social conservatives to call for an end to mixed-gender training at the basic level. With the collapse of the Saddam Hussein regime in April 2003, the prison twenty miles west of Baghdad was renovated and became a U.S. military prison. In late February 2004, the conclusions of a report on the prison written by Maj. Gen. Antonio M. Taguba were devastating. The report revealed institutional failures and numerous instances of "sadistic, blatant, and wanton criminal abuses" against prisoners by soldiers of the 372d Military Police Company, some private contractors, and members

---

[53] A front page article in the *Washington Post*, dated 3 Apr 2003, portrayed Lynch as a "woman warrior" who was shot, stabbed, and captured only after emptying her weapon killing Iraqis. A clarification was issued in the 20 Apr 2003 edition.

[54] Kathleen Parker, "Jessica Lynch's Story Is About a Girl, Not a Soldier," *Townhall*, 19 Nov 2003.

of the American intelligence community. Photographs and videotapes taken by the soldiers—and later broadcast on television—provided stunning evidence of Taguba's allegations. Charges against the soldiers, who included three women, ranged from conspiracy, dereliction of duty, and cruelty toward prisoners to maltreatment, assault, and indecent acts.[55]

The incidents at Abu Ghraib sparked additional publication of articles by conservatives, who blamed the Army's "politically correct" basic training program:

> The new politically correct way of training recruits is failing to impart a sense of discipline, and by all accounts the mistreatment of prisoners at Abu Ghraib reflected more than anything else a breakdown of discipline. This is perhaps the most serious—and most important—lesson we can draw from this sorry episode....The Abu Ghraib "horror" has revealed once and for all an even greater horror: those who advocate the sexual integration of all aspects of society, the removal of all sexual taboos, and the feminization of traditionally masculine worlds like boot camp—are winning....It's no wonder that some of our service-*persons* would turn a combat-zone prison into a theater of sexual performance art.[56]

Another conservative columnist adopted a somewhat different theme: "The pictures from Abu Ghraib of male Iraqi prisoners stripped naked, demeaned, and sexually humiliated by female soldiers blows the lid off the theory of a kinder, gentler military....Could it be that all this gender-norming has increased the pressure on military women to conform to the behavior patterns of the men around them in order to prove they belong, even when they know this behavior is wrong?"[57]

Phyllis Schlafly, president and founder of the Eagle Forum, called the incidents at Abu Ghraib a public relations disaster. She continued, "Just as humiliating for Americans is allowing the world to see the depths reached by a gender-integrated U.S. military....The pictures are stark illustrations of the gender experimentation that has been going on in the U.S. military. The images have lifted the curtain on a subject about which the public has largely been kept in the dark."[58]

---

[55] Seymour M. Hersh, "Torture at Abu Ghraib: American Soldiers Brutalized Iraqis. How Far Up Does the Responsibility Go?" *The New Yorker*, Vol. 80, No. 11, 10 May 2004.

[56] Rev. Aris P. Metrakos, "S&M Soldiers: How Sexual Politics Has Undermined the U.S. Military," *Orthodoxy Today*, 2004. Rev. Metrakos is pastor of Holy Trinity Greek Orthodox Church in Columbia, S.C., home of Fort Jackson.

[57] Jane Chastain, "So Much for a Kinder, Gentler Military," *World Net Daily*, 6 May 2004.

[58] "Feminist Dream of Military Equality Becomes Nightmare in Iraq," Copley News Service, 17 May 2004.

Conservative columnist Cal Thomas observed: "The one dirty little secret that no one appears interested in discussing as a contributing factor to the whorehouse behavior at Abu Ghraib is coed basic training and what it has done to upset order and discipline."[59]

But retired Navy Captain Lory Manning wrote that the photographs of Abu Ghraib "had nothing to do with gender. They show only that women are capable of making the same mistakes as men."[60]

As of the early fall of 2004, the courts-martial of the prison guards continued, and there were at least a dozen investigations—some ongoing. Preliminary findings indicated that a lack of leadership and a lack of training in handling prisoners were primary causes of the incidents at Abu Ghraib and other Army prisons. Not nearly resolved was the question of whether the involvement of women at Abu Ghraib would affect the continuing debate about the role of women in America's military.

A New BCT Program of Instruction: Warrior Ethos

The fighting in Iraq, a battlefield without borders, was representative of a return to an old form of irregular warfare. As such, it demanded a new POI for basic training of recruits. Opponents of mixed-gender training in BCT once again argued that the combat exclusion laws still stood, and thus, there was no reason to retain integrated training. In reality, the public could clearly see and generally accepted that women were working alongside the war fighters, taking hostile fire—even in the role of designated support forces. For the Army, 90 percent of MOSs were represented either in Iraq or the continental United States. The consequences of a global war on terror extended beyond military job skills. For example, women serving as Military Police often had the same mission as all-male combat units, and female soldiers made up 25 percent of the Military Police Corps.[61] As 2004 ended, more women had been killed and wounded in Iraq and Afghanistan than in any conflict since World War II.[62] As Capt. Lory Manning put it, "[women] aren't safe anywhere, in combat or otherwise, and it's not news anymore."[63]

[59] "Sexual Politics and the Breakdown at Abu Ghraib," *Baltimore Sun*, 19 May 2004.

[60] Darryl Fears, "Military Families Mourn Daughters: 20 Female Service Members Have Been Killed in Iraq," *Washington Post*, 26 May 2004, pp. A-1, A-19.

[61] Women made up approximately 16 percent of Army forces overall. One Station Unit Training (OSUT) was predominately all male. However, OSUT for Military Police was fully integrated.

[62] Monica Davey, "For 1,000 Troops, There Is No Going Home," *New York Times*, 9 Sep 2004, pp. A-1, 20–24.

[63] Tranette Ledford, "Women at War," *Army Times*, 24 Nov 2003.

Thus, when TRADOC's trainers began to redesign basic training, arguments for gender separation seemed groundless.

In late 2003, Army Chief of Staff Gen. Peter J. Schoomaker called for the establishment of an IET Task Force at TRADOC headquarters to study and make recommendations for changes in BCT to prepare basic trainees for the insurgency and guerrilla warfare taking place in Iraq and Afghanistan.[64] Prior to this study, recruits received no drills in the prevention of damage from weapons, such as rocket propelled grenades and improvised explosive devices, nor was instruction provided in conducting convoys, patrolling, manning checkpoints, or countering urban warfare tactics. The Army's basic training was still geared toward fighting a conventional war in which front lines were established and soldiers fought in large units according to a prescribed set of orders and procedures. General Schoomaker expressed his belief that it was time to incorporate the lessons learned in Iraq and Afghanistan into a new training system. His decision, by his own admission, was informed by the experience of the 507th Maintenance Company, which lost eleven soldiers during an ambush in Iraq in March 2003. An Army investigation found that many of the soldiers of the 507th were unable to defend themselves because their weapons malfunctioned and they were outmaneuvered by Iraqi irregulars (see above).[65]

The changes that the chief of staff directed the task force to examine would be the greatest since the Vietnam War, perhaps since World War II. Changes included more weapons training and instruction in firing weapons other than the M-16, including a variety of machine guns. Recruits would be taught how to identify and counter remote-controlled bombs. For the first time, trainees would ride in convoys and face simulated ambushes. They would be taught how to place sandbags inside vehicles as protection against bombs and grenades. Soldiers would learn how to fight in urban areas where enemies blended in with civilians. First-aid training would be significantly increased to include more lifesaving skills, because troops traveling in small groups could be ambushed without a medic or doctor available. Schoomaker surmised: "Transforming IET is about breaking contact with a system designed

---

[64] Primary responsibility rested with the relatively new U.S. Army Accessions Command, which had been activated 25 Mar 2002, at Fort Monroe, Va. The new command was part of TRADOC. The tasking to the new task force also included changes to AIT. Early in FY 2004, Basic Combat Training became known as Initial Military Training. For consistency, the author has retained the earlier term.

[65] Chuck Crumbo, "Army May Revise Basic Training at Fort Jackson," www. thestate.com, South Carolina's Home Page, 7 Jan 2004; Moniz, Dave, "Army Revamps Training: Insurgent Tactics Prompt Changes," *USA Today*, 6 Jan 2004.

in the 1960s for a draft era, CAT IV Army that no longer exists and a World War that never took place....Success or failure hinges on fundamentally changing the nature, character, and conduct of the relationship between America's volunteers and their first NCOs."[66]

Perhaps most important, the strategy for the future would include a warrior ethos and would incorporate a soldier's creed. Tenets of the warrior culture supported this new approach: mission first; never accept defeat; never quit; and never leave a fallen comrade behind. The soldier's creed included the definition of *warrior ethos* and stressed commitment, teamwork, discipline, and professionalism. Soldiers whose jobs would have traditionally kept them far from the front lines—cooks, clerks, and truck drivers—would also undergo combat training. In short, everyone would be a warrior, not just those in the combat arms MOSs. Army Chief of Staff Schoomaker believed that soldiers needed to "refocus their attention on what it means to be a warrior instead of focusing as much on their military specialty."[67] A new basic training program would feature many additional "warrior tasks," as well as more field training exercises (up to sixteen days versus less than a week). One component that was not negotiable as a pillar of a new training system were the Army values, adopted shortly after the sexual misconduct allegations at Aberdeen Proving Ground in 1996–1997 (see Chapter IV). Also nonnegotiable was gender-integrated training for soldiers in combat support and combat service support MOSs.[68]

On 27 February 2004, the IET Task Force presented an in-process review (IPR) for TRADOC commander Gen. Kevin Byrnes based on three ongoing pilot BCT programs (January–May 2004).[69] The task force recommended, among other things, that the ratio of "leader to led" (drill sergeant to recruit) be lowered to 1:10. As one general officer put it, "field time is a 'camp-out' without more [command] personnel." To accomplish a 1:10 ratio without increasing NCO strength, it might be necessary to eliminate drill sergeants in advanced individual training (AIT), replacing them with platoon sergeants and contractors, and that One Station Unit Training (OSUT) train only MOSs 19D (cavalry scout) and 11B/C (infantryman/indirect-fire infantryman). At that time, estimates were based on forty individual warrior tasks and nine warrior battle drills. The IET Task Force suggested but did not recommend that some BCT tasks not requiring a 1:10 ratio of drill sergeants to trainees

---

[66] Initial Entry Training Task Force, in-process review to commanding general, TRADOC, 27 Feb 2004, p. 3.
[67] TRADOC News Service, TRADOC Public Affairs.
[68] Initial Entry Training Task Force.
[69] The pilot programs were conducted at Forts Benning, Knox, and Jackson.

be moved into the recruiting or reception phases. The panel also offered General Byrnes comparisons of three BCT programs of instruction: eight-week BCT; a revised nine-week BCT; and ten-week BCT. The task force also compared three options for mixed-gender training: (1) no gender-integrated training (GIT); (2) all BCT as mixed-gender; and (3) maintenance of the status quo of GIT in programs where combat exclusion was not an issue.[70] The task force recommended the third option.

In June 2004, the ongoing changes in BCT, as suggested by lessons learned programs in Iraq and Afghanistan, were endorsed at a meeting of the Army's training brigade commanders and put into effect on an interim basis at all five basic training sites. Formal acceptance of a combat-oriented BCT awaited General Schoomaker's approval. In addition, increased funding was necessary for realistic live-fire training and longer field training exercises. Meanwhile, the task force and the senior Army leadership were finding that the "IET assembly line mentality [is] hard to break."

On 27 August 2004, a General Officer Steering Committee presented the IET Task Force's recommendations, as amended since the February IPR, to General Byrnes. The TRADOC commander did not support removing all drill sergeants from AIT; in fact, their responsibilities would increase as they reinforced the warrior ethos and battle drills training taught in BCT. Likewise, he did not support limiting OSUT to only two combat arms MOSs. BCT would remain a nine-week POI, and there would be no personnel increases. The initial recommendation for forty warrior tasks and nine battle drills was reduced to thirty-two tasks and five drills. The requirement would remain, but time and resource limitations dictated that all would not be taught in BCT. Gender-integrated training in BCT remained in place. In the immediate aftermath of Byrnes's decisions, it was not clear what if any effect the new BCT POI would have on mixed-gender training. Based on this, however, it could be assumed that opponents would continue to protest gender-integrated basic training as a compromise to military readiness, regardless of changes.[71]

A New Physical Training Program

Men and women soldiers in mixed-gender BCT would also experience a new physical training program designed not only to make train-

---

[70] Initial Entry Training Task Force, 27 Feb 2004, pp. 7, 13, 27, 28–30, 40.

[71] Initial Entry Training Task Force review, General Officer Steering Committee recommendations, 27 Feb 2004.

ing of combat-based tasks more efficient and effective but to ensure success on the Army physical fitness test and the deployment of well-conditioned soldiers. It was also hoped that the new program would lower rates of injury, thereby reducing attrition. A study at Fort Jackson found that 50 percent of women and 25 percent of men suffered some injury, especially stress fractures, during the nine-week training period. The high rates of injury and attrition among female trainees had long been used by opponents of gender-integrated basic training as an argument for abandoning the program.[72]

Beginning on 1 April 2004, the TRADOC-directed IET standardized physical training program was implemented at all initial-entry locations. The POI was based on studies and reviews by a number of experts in physiology, fitness, and sports medicine and was developed at the U.S. Army Physical Fitness School at Fort Benning, Ga. The revised guides for IET outlined a two-phase schedule of physical training activities designed to challenge soldiers entering the Army regardless of gender, age, or fitness level. The toughening phase was designed to develop fitness and fundamental skills. Soldiers learned to manage body weight. Through calisthenics and movement drills, soldiers learned skills essential to the battlefield, such as jumping, landing, lunging, bending, reaching, and lifting. Especially important for women were exercises that targeted the muscles in the lower extremities to improve flexibility, strength, and endurance. The training events in this first phase were designed to ensure that bones, muscles, and connective tissue gradually toughened rather than broke. The conditioning phase was designed to develop a high level of physical readiness appropriate to duty position. Soldiers in BCT were in the toughening phase throughout most of their training, usually not entering the conditioning phase until AIT.[73] Recruits who failed to meet unit goals or Army standards would have available special conditioning programs, that is, remedial programs designed to meet individual needs to overcome identified weaknesses. The aim of the increased emphasis on physical training was to ensure that all soldiers—men and women—were physically prepared for the demands of deployment, which would likely come earlier in their careers than in years past.[74]

The new programs in BCT and AIT dictated changes in the POI in the three drill sergeant schools, because drill sergeants were solely

---

[72] Capt. Angela Hildebrant, U.S. Army Accessions Command, TRADOC News Service, 26 Feb 2004; Crumbo, 20 Jun 2004, www.thestate.com.

[73] U.S. Army Basic Training Study Guide, 2004.

[74] Memo for distr, ATAL-O, Maj Gen Russell Honoré, "IET Assessment Recommendations and Decisions," Jul 2004.

responsible for conducting BCT. The warrior ethos concepts were incorporated into the instruction, as was the new physical training program. In addition, candidates for the position of drill sergeant learned how to instruct privates in the combat-related skills then becoming part of a recruit's training. In another change, drill sergeants would act as squad leaders during field training exercises. In a related action, TRADOC commander General Byrnes directed that drill sergeants not be employed for installation support. "Drill sergeants are trained to train and lead soldiers. That's what I want them doing."[75]

Women and the Draft

When months of insurgency followed the fall of Saddam Hussein's government in Iraq and peacekeeping in Afghanistan continued, U.S. and Coalition Forces—especially the Army—were stretched thin. It was, perhaps, inevitable that the already much debated subject of reinstating Selective Service—the draft—would resurface. But what about women and the draft? Should women, like men, be required to register? Should anybody be required to register? Should women actually be drafted? Should combat assignments be voluntary for women? What would a draft mean for combat exclusion? At this writing, the debate continues, as does the demand for more ground troops and the possibility of more operations like the one in Iraq. As with the question of training men and women together, the draft question tended to polarize debate.

Those who are opposed to drafting women used arguments identical to those employed by opponents of gender-integrated basic training. Opponents argued that women would spoil esprit de corps and unit cohesion; women weren't physically strong enough; women would distract men from their mission; women would be vulnerable to rape and assault as POWs; the public was not ready to see women in body bags; and men would lose their lives trying to protect women. One conservative columnist summed up the argument: "Yet how are these women to survive combat if they cannot survive real, not gender-normed, basic training? The men would have to protect them. Successfully integrating women in combat means this: A soldier must ignore the screams of a woman POW being tortured and raped."[76]

Others argued that the purpose of a draft was to create a combat-ready force, and because legally, women weren't allowed in direct

---

[75] 18 Aug 2004, www.thestate.com.
[76] R. Cort Kirkwood, "What Kind of Nation Sends Women into Combat?" *Daily News-Record*, Harrisonburg, Va., 11 Apr 2003.

ground combat, a female draft was irrelevant. In short, registration for a draft and the draft itself remained tied to women's exemption from direct ground combat and vice versa.[77]

The inclusion of women in the conscription pool also had its defenders—for a variety of reasons. Many men thought that if women desired equal treatment and equal opportunity, such also ought to apply to military service. Many women, perhaps a majority, thought that women should be drafted but that service in combat arms units should be optional. Liberal columnist Anna Quindlen, writing in *Newsweek*, believed that young women should be required to sign up and called the failure to send women into combat "unfair."

Sociology professor and military analyst Charles Moskos suggested a three-tier draft system to include national service: combat, homeland duty, and civil service. Women would be registered just as men were, but they would be drafted only for tiers two and three.[78] Although his topic did not concern women, Rep. Charles B. Rangel (D-N.Y.) received considerable attention from those who opposed the draft for women when he proposed that the draft be reinstated to eliminate what he saw as a disproportionate burden on the poor and minorities.[79]

Perhaps the most visible event regarding women and the draft was a suit filed in U.S. District Court in Boston in January 2003. Five students (four men and one woman) challenged the constitutionality of military registration because it required that only men register. Men, it alleged, were being denied the rights of due process and equal protection under the Fifth and Fourteenth amendments. Essentially, the suit challenged legislation passed by Congress in 1980 that excluded women from registering. The law was enacted when President Jimmy Carter reinstated mandatory registration and argued women should be included (see Chapter III). In 1981, the law was challenged, and the U.S. Supreme Court affirmed the constitutionality of the male-only draft registration system and upheld the right of Congress to exclude women from Selective Service registration (*Rostker* v. *Goldberg*). The students argued that the courts should reconsider the ruling because the acceptance of women in the military had progressed so far that there was no justification for men having to bear the entire burden of war. The Department of Justice, defending the Selective Service System, argued that it was not for the courts to judge but was, rather, a congressional

---

[77] Maura Jane Farrelly, "Iraq War Renews Women in Combat Debate," *Iraq Crisis Bulletin*, 1 Apr 2003.

[78] Joan Ryan, "Women and Uncle Sam," *San Francisco Chronicle*, 14 Jan 2003.

[79] Tom Infield, "Today's Army Differs from Gulf War Force," *Philadelphia Inquirer*, 2 Feb 2003.

prerogative.[80] On 29 May 2003, the district judge upheld the 1981 ruling, declaring that the judiciary lacked the power to make policy in this arena, as follows: "If a deeply rooted military tradition of male-only draft registration is to be ended, it should be accomplished by that branch of government which has the constitutional power to do so, and which best represents the 'consent of the governed'—the Congress of the United States, the elected representatives of the people."[81]

However, as long as the war in Iraq continues and manpower needs continue to increase, the question of registering or conscripting both men and women would continue to be debated in public and private forums.

Capt. Lory Manning USN (Ret.), director of the Women in the Military Project, perhaps put it best: "Unless we want to draft men, we have to take women. We have to have an all-volunteer service or go back to the draft. That's the trade-off."[82]

---

[80] Thanassis Cambanis, "U.S. Contests Lawsuit vs. Draft: Argues Against Including Women," *Boston Globe*, 20 May 2003, p. B-3.

[81] *Samuel Schwartz et al.* v. *Lewis C. Brodsky*, Director of the U.S. Selective Service, and Attorney General Thomas Reilly of Massachusetts, Civil Action No. 03-10005-EFH, 29 May 2003.

[82] Farrelly, 1 Apr 2003.

Conclusion

● ●

Early in 1973, the United States Army became an all-volunteer force (AVF) when the conscription (draft) of young men into its ranks ended. In order to maintain the desired numbers on active duty, the Army found it necessary to recruit increasingly large numbers of young women to serve alongside the traditional all-male force. In turn, the service had to define the role of this larger number of women and the mode of training they would undergo. Should men and women be trained together, especially in basic combat skills, for the "new" Army? In its efforts to define and execute its policies toward women soldiers, the Army encountered an increasingly polarized debate expressed in political, social, economic, and military terms. At every turn, from the competing points of views inside and outside the Army on the role of women and their training for assignments, it became clear that the Army is a microcosm of the society in which it operates and the political system by which it is sustained.

In addition and, perhaps, most important, the AVF took form concurrently with a strong women's rights movement in the United States. Although most American feminist groups paid little attention to the issues of women in the military and sometimes condemned women's participation in the military, the movement's existence pushed the Army to clearly formulate its policies concerning women and, in particular, enlisted female soldiers. Even though the effort ultimately failed, the proposed Equal Rights Amendment of 1972 contributed more than anything to the unprecedented expansion of women's participation in the American armed forces. The Army's policies, in turn, tended to drive how congressional conservatives, women's equal rights organizations (pro and con), and the legal system reacted to the integration of women into the military services, especially the Army, with its preponderance of ground forces.

A number of issues, some of which had long been concerns of women soldiers and the Army, provided the background for and informed the debate about the employment and training of an increasingly large number of female enlistees and the training of units that included both sexes during their initial introduction to the service. At the same time, the existing social, military, and political climate went far in defining the issues the Army had to consider in its effort to design a mixed-gender training program for basic combat training (BCT).

As the proportion of women in military service rose, numbers alone caused great uncertainty and misgivings in a historically male institution concerning military readiness and mission capability. The training base and supply system seemed unprepared for the greatly increased numbers of women who required training. By 1998, one in five enlisted recruits was female.

At the same time, the interrelated questions of the draft, women in combat, and the role of women in the U.S. armed forces cast a long shadow over the debate concerning what the nature of mixed-gender basic training should be. Many critics of the AVF saw reinstatement of the draft for men as a means of reducing dependence on female military personnel. Concurrently, conscription for both men and women was publicly unpopular. Meanwhile, a national consensus on the proper role of women in national defense was further hindered by the exclusion of women from combat. In 1980, the Carter administration's efforts to reinstate draft registration for men and women were defeated by congressional armed services committees on the grounds that women were excluded from combat.

The entire debate was further inflamed by the absence of an adequate definition of *combat*. For the Army, as more women joined the force and were allowed in a greater number of military occupations, defining where women could serve and how they should be trained depended on an ever-changing definition of *combat*. The term had traditionally been defined in terms of a unit's relation to the enemy on the battlefield. That definition lost its validity in the presence of advancing technology. At the same time, women's rights groups continued somewhat hesitantly to insist that women would not have full citizenship until they were subject to the draft. As long as the issues of women's assignment to combat roles and the draft remained unresolved, the Army found design of a new training program difficult.

In addition to military controversies, other issues consistently set the parameters for the debate about the role of women in the military and provided the background for the Army's efforts to define the structure of basic combat training for the AVF. Divisions of opinion—subject to change within the military, social, and political climate—were not always between men and women, nor was there often agreement among the various branches of the government or the military.

Perceived differences between men and women in physical strength and endurance were most influential in shaping the debate surrounding basic training in mixed-gender units. Were the physiological differences in strength, stamina, speed, and coordination genetically determined or the product of a less active culture among women and, therefore, subject to change? The Army conducted a multitude of tests in an effort to answer this question. Others argued that advancing technology would decrease the importance of strength and endurance. Still others feared that compromises would threaten military readiness. In support of this view, male trainers often asserted that integrated training for men and women compromised their own training.

Close behind physical fitness standards as the most debated issues concerning the role of women in the military were marriage, pregnancy, and parenthood. Initially, for a young woman, military service was seen as a temporary career until marriage. There was no thought at all that she might become a mother. In the days of the Women's Army Corps (WAC), marriage or pregnancy meant mandatory discharge. By the 1960s, married women could join any of the services. The advent of the AVF caused the Department of Defense to rethink pregnancy policy. In the mid-1970s, separation for pregnancy became voluntary, but concerns for under-strength units, deployability, readiness, and mission accomplishment remained. These same concerns seemed to apply to single parents of either sex and to couples with both partners in the service. All these issues came together during Operation DESERT SHIELD and Operation DESERT STORM in 1990–1991 when new questions about deployment of single parents and dual-service couples surfaced. The parenting issue had become a byproduct of the AVF.

Discussions of the relative conditions of military service for women and men inevitably turned to attrition and retention and their effects on readiness and cohesion. Concerns about severe losses of enlisted women predated the AVF. Beginning in the 1950s, various Department of Defense policy changes sought to prevent the use of marriage as justification for the separation of women before completion of their enlistment contracts. Despite those changes, during the Cold War of the 1950s and 1960s, 70 to 80 percent of female recruits left military service before completing their first enlistment. That phenomenon meant that the replacement rate for women was two and one-half times that for men. With the advent of the AVF, senior leaders increasingly believed that more women would be necessary to fill the ranks as the numbers of military women rose from 2 percent to 10 percent. High attrition rates became alarming and served as powerful ammunition for those opposed to women in the military. Some argued that high replacement rates wasted the training investment. Defenders of women in the military often argued that women's exclusion from combat frustrated their ambitions for promotion and contributed to the high attrition rates.

An issue that received much publicity and created deep concern for the military leadership was the potential adverse impact of the presence of women in formerly all-male units, particularly on the desire of men to maintain a masculine warrior ethic. Would the presence of women disrupt men's interpersonal relationships and crucial bonding and cohesion among male soldiers? Could men remain the protectors and women the protected? How were manhood and masculinity to be validated if women were able to perform traditionally male jobs successfully? Opponents and proponents of mixed units found one point of agreement—social justice as a national objective had to be considered alongside military preparedness.

Central to the debate about mixed units for basic training was the belief inside and outside the military services that there were significant behavioral differences between men and women and in their levels of aggression and fighting spirit. Perhaps, some argued, these traits could be determined or changed during the socialization process. Also debatable was the question of whether modern warriors needed to be as aggressive as those in the past. Further, were the most aggressive individuals likely to be the most disciplined soldiers? Those opposed to mixed-gender units held that women were less combative than men and likely to have a negative effect on male fighting performance. Proponents held that women would not have full equality until they served equally with men.

Three other issues also acted to provide the background against which the Army sought to determine the makeup of the units that would train new enlistees in basic combat skills, especially given the increasing number of female recruits in the AVF. Fraternization (close personal relations between officers and enlisted personnel or seniors and their subordinates) was certain to be an issue in the training of young men and women in mixed units. Generally, the U.S. Army tended to leave such matters to local commanders—that is, until the late 1970s, when the first of a series of regulations defined relationships likely to compromise impartiality or to undermine morale, discipline, or authority. Disciplinary authority remained with commanders according to the Uniform Code of Military Justice. A follow-on regulation of 1999 expressly prohibited relationships between career service members and initial-entry trainees.

As the percentage of women in the Army grew from 2 percent in the early 1970s to 15 percent at the beginning of the twenty-first century, sexual harassment caused increasing concern for the military leadership, soldiers, officers, and some focus groups in the general public. Not the least of the problem, for all the military services, was defining *harassment*. Also of concern was determining at what point tough and demanding training became abuse. Although Army senior leaders recognized that the fact of more women in the ranks was likely to result in more incidents of harassment, the service dealt with the problem on a case-by-case basis while supporting educational programs. In the fall of 1996, the arrest of numerous soldiers at Aberdeen Proving Ground in Maryland and at other installations for sex crimes prompted a media frenzy and congressional hearings. The secretary of the Army and the secretary of Defense each appointed commissions to review military training, especially the training of men and women in mixed-gender units. Congress also created a commission. These panels addressed a variety of issues with a variety of results. Only one thing was clear— sexual harassment continued to defy definition.

Of major importance to the definition of women's role in the armed forces and the future of mixed-gender basic training was public opinion, the so-called "will of the American people," which had far-reaching implications for the attitudes and actions of members of Congress, the courts, and military leadership. Public attitudes, despite numerous polls and surveys, proved impossible to define. Generally, the public supported increasing opportunities for women in the military to parallel those for women in the civilian sector. At the same time, the public will was often invoked to justify exclusion of women from combat and the draft.

Having determined that men and women who would likely fight together should be trained in basic combat skills together, the Army designed and executed a program of mixed-gender training from 1978 to 1982. That program was abandoned after four years. After the failure of this initial mixed-gender training program, public forums became increasingly polarized on the issue of the reinstitution of gender-integrated training. That debate ended in 1994 with a new gender-integrated basic training program in the wake of what many saw as the success of women soldiers in Operation DESERT SHIELD and Operation DESERT STORM. During the next two years, a new gender-integrated training program addressed and resolved a number of issues, while the question of women in combat still awaited resolution. In 1996, a number of reported incidents of sexual harassment and abuse in gender-integrated units brought renewed criticism of mixed-gender training and a series of congressional studies. Generally, the congressional reports received little attention and the debate about gender-integrated training at the basic level once again dwindled, except in the media that had traditionally supported gender-pure training. The criticism there was, once again, based on opposition to social policies and social experimentation, which some believed threatened to negatively affect military readiness and national security. The Army's senior trainers in the Training and Doctrine Command (TRADOC) and the Forces Command (FORSCOM) met repeated attempts by some members of Congress and antifeminist groups to abolish the gender-integrated program with arguments that mixed-gender basic training was effective if not always cost-efficient. The terrorist attacks on U.S. sites of 11 September 2001 tended to turn attention away from training programs for women recruits.

Pressure on President George W. Bush's administration to direct the military services to return to gender-segregated basic training took the form of a movement to abolish the fifty-year-old Defense Advisory Committee on Women in the Services (DACOWITS), which had long supported expansion of opportunities for female soldiers and changes in policies excluding women from combat. The Bush administration's "hands-off" posture resulted in a new and severely limited charter for DACOWITS.

In late 2004, as the war continued in Iraq, the Army drastically revised its program of instruction for BCT to incorporate lessons learned in Iraq and Afghanistan and to meet the challenges of irregular warfare, insurgency, and urban tactics. In so doing, the service abandoned the training system that had been in place since the 1960s and had been designed for a draft era. Gender-integrated training continued and mixed-gender basic training units remained the basic component of initial-entry training. At the same time, opponents of gender-integrated training continued their efforts to end a training system that they saw as the creation of radical feminists and that threatened to sacrifice national security on the alter of political correctness.

At this writing, the large number of female soldiers serving in mixed-gender units in Iraq and the changing face of modern warfare seems certain to bring to the fore, once again, the question of gender-integrated training of men and women in basic combat skills during initial-entry training.

Selected Bibliography

• •

Unless otherwise indicated, all primary sources are located in the documents collection at the U.S. Army Training and Doctrine Command Military History Office, Fort Monroe, Va. Copies of many of the secondary volumes and articles are also located there. No attempt has been made to record URLs for Internet sites because they change often. Some information from the Internet is available in hard copy at the Military History Office.

U.S. Army and Department of Defense Documents

Army Regulation 600–20. Army Command Policy, 5 March 1993.

Berrong, Larry B. "A Case for Women in Combat." U.S. Army Command and General Staff College, June 1977.

Blank, Stephen J. "Rethinking Asymmetric Threats." Strategic Studies Institute, Carlisle Barracks, Pa., September 2003.

Department of the Army. Human Relations Action Plan, "The Human Dimensions of Combat Readiness," September 1997.

Hart, Roxine C. "Women in Combat," Defense Equal Opportunity Management Institute, 1990.

Hays, Mary Sue, and Charles G. Middlestead. "Women in Combat: An Overview of the Implications for Recruiting," Alexandria, Va.: U.S. Army Research Institute for the Behavioral Sciences, July 1990.

Jeffrey N. McNally. "Women in the United States Military: A Contemporary Perspective," Newport, R.I.: Naval War College, 1985.

Lipari, Rachel N., and Anita R. Lancaster. "Armed Forces 2002 Sexual Harassment Survey," Defense Manpower Data Center, February 2004.

Mottern, Jacqueline A., et al. "Integration of Basic Combat Training Study," Alexandria, Va.: U.S. Army Research Institute for the Behavioral Sciences, February.

Myers, David C., Deborah L. Gebhardt, Carolyn E. Crump, and Edwin A. Fleishman. "Validation of the Military Entrance Physical Strength Capacity Test," Technical Report 610, Alexandria, Va.: U.S. Army Research Institute for the Behavioral Sciences, January 1984.

Nogami, Glenda Y. "Fact Sheet: Soldier Gender on First-Tour Attrition," Alexandria, Va.: U.S. Army Research Institute for the Behavioral Sciences, 1981.

Office of the Assistant Secretary of Defense for Manpower, Reserve Affairs, and Logistics. "Use of Women in the Military," Washington, D.C.: May 1977.

"Performance on Selected Candidate Screening Test Procedures Before and After Army Basic and Advanced Individual Training," Natick, Mass.: U.S. Army Research Institute of Environmental Medicine, June 1985.

Report of the Secretary of the Army's Senior Review Panel on Sexual Harassment, July 1997.

Report of the Federal Advisory Committee on Gender-Integrated Training and Related Issues to the Secretary of Defense, Executive Summary, 16 December 1997.

Savell, Joel M., Carlos K. Rigby, and Andrew A. Zbikowski. "An Investigation of Lost Time and Utilization in a Sample of First-Term Male and Female Soldiers," Technical Report 607, Alexandria, Va.: U.S. Army Research Institute for the Behavioral Sciences, October 1982.

U.S. Army Administration Center. *Evaluation of Women in the Army,* Final Report, March 1978.

U.S. Army Audit Agency. "Enlisted Women in the Army," April 1982.

U.S. Army Basic Training Study Guide, 2004.

U.S. Army News Service. Department of the Army Public Affairs, 30 July 1998; 2 March 1999.

U.S. Army Regulation 600-20. *Army Command Policy,* November 1978 and subsequent editions.

U.S. Army Research Institute. Qualitative Data Collection Component of Gender Integration of Basic Entry Training Study, Fort Jackson, S.C., 1993–1994 [unpublished].

U.S. Army Training and Doctrine Command. Annual Historical Reviews [also titled Annual Historical Summary and Annual Command History], FY 1975–FY 1994.

U.S. Army Training and Doctrine Command. Initial-Entry Training Review Task Force, In-Process Review to Commanding General, 27 February 2004.

U.S. Army Training and Doctrine Command. TRADOC Regulation 350-6, November 1998; July 2001; October 2003; C1 to TR 350-6, August 2003.

U.S. Army Training Center at Fort Jackson, S.C. "Gender-Integrated Training," Briefing to Secretary of the Army, 22 March 2002.

U.S. Department of Defense. Office of the Secretary of Defense. "Report of Task Force on Women in the Military," January 1988.

U.S. Department of Defense. *Selected Manpower Statistics,* Office of the Assistant Secretary of Defense, Directorate for Information, Operation, and Control, May 1975.

U.S. Department of the Army Inspector General. "Special Inspection of Initial Entry Training Equal Opportunity/Sexual Harassment Policies and Procedures," 22 July 1997.

U.S. Department of the Army. Historical Summary, FY 1975–FY 1991. Washington, D.C.: Center of Military History.

———. "Program of Instruction for Basic Training (BT) of Male and Female Military Personnel Without Prior Service (Six Weeks), December 1975" [POI 21-114 Test Edition].

———. Office of the Deputy Chief of Staff for Personnel. *Women in the Army Study.* Washington, D.C.: December 1976.

U.S. Department of the Army. Office of the Deputy Chief of Staff for Personnel, *Women in the Army Policy Review,* 12 November 1982.

"Women Content in the Army–REFORGER 77," Special Report S-7, Alexandria, Va.: U.S. Army Research Institute for the Behavioral Sciences, 30 May 1978.

"Women Content in Units: Force Development Test (MAX WAC)," Alexandria, Va.: U.S. Army Research Institute for the Behavioral Sciences, 3 October 1977.

Public Documents

Department of Defense, *Selected Manpower Statistics,* Office of the Assistant Secretary of Defense. Directorate for Information, Operation, and Control, May 1975.

Statutes at Large 95-202, sec. 401 (1977).

U.S. Code, Title 10, Sec. 3012.

U.S. Congress, House of Representatives, House Armed Services Committee. Initial Report of the Congressional Committee on Military Training and Gender-Related Issues (Blair Commission). 106th Cong., 1st sess., 17 March 1999; 31 July 1999.

U.S. Congress, House of Representatives. *Sexual Harassment in the Federal Government*, 96th Cong., 2d sess., 30 April 1980.
U.S. Congress, Senate, Subcommittee on Defense on the FY 1984 Army Budget Overview. Hearing before the Senate Appropriations Committee. 98th Cong., 1st sess., 24 February 1983.
U.S. Government Accounting Office. "Gender Issues: Analysis of Methodologies in Reports to the Secretaries of Defense and the Army," Washington, D.C., 16 March 1998.
———. "Job Opportunities for Women in the Military: Progress and Problems," GAO/FPCD-76-26, May 1976.
———. "Women in the Military: Impact of Proposed Legislation to Open More Combat Support Positions and Units to Women," July 1988.
———. "Women in the Military: More Military Jobs Can be Opened Under Current Statutes," September 1988.
———. "Women in the Military: Deployment in the Persian Gulf War," 13 July 1993.
Women's Armed Services Integration Act of 1948, 62 Stat. 356–75.

Secondary Sources—Books

Binkin, Martin, and Mark J. Eitelberg. "Women and Minorities in the All-Volunteer Force." In *The All-Volunteer Force After a Decade: Retrospect and Prospect.* Edited by William Bowman, Roger Little, and G. Thomas Sicilia. Washington, D.C.: Pergamon-Brassey's, 1986.
Binkin, Martin, and Shirley J. Bach. *Women and the Military.* Washington, D.C.: The Brookings Institution, 1977.
Davis, Flora. *Moving the Mountain: The Women's Movement in America Since 1960.* New York: Simon & Schuster, 1991.
Enloe, Cynthia H. "The Politics of Constructing the American Woman Soldier." In *Women Soldiers: Images and Realities.* Edited by Elisabetta Addis, Valeria E. Russo, and Lorenza Sebesta. New York: St. Martin's Press, 1994, ch. 5.
Fowler, William W. *Frontier Women: An Authentic History of the Courage and Trials of the Pioneer Heroines of Our American Frontier.* Stamford: Longmeadow Press, 1995.
Francke, Linda Bird. *Ground Zero: The Gender Wars in the Military.* New York: Simon & Schuster, 1997.
Friedl, Vicki L., comp. *Women in the United States Military, 1901–1995: A Research Guide and Annotated Bibliography.* Westport, Conn.: Greenwood Press, 1996.

Fullinwider, Robert K., ed. *Conscripts and Volunteers: Military Requirements, Social Justice, and the All-Volunteer Force*, Maryland Studies in Public Philosophy. Rowman & Allanheld, 1983.

Goldman, Nancy Loring, ed. *Female Soldiers—Combatants or Noncombatants: Historical and Contemporary Perspectives.* Westport, Conn.: Greenwood Press, 1982.

Gutmann, Stephanie. *A Kinder, Gentler Military: Can America's Gender-Neutral Fighting Force Still Win Wars?* New York: Scribner, March 2000.

Herbert, Melissa S. *Camouflage Isn't Only for Combat: Gender, Sexuality, and Women in the Military.* New York: New York University Press, 1998.

Holm, Jeanne. *Women in the Military: An Unfinished Revolution*, rev. ed. Novato, Calif.: Presidio Press, 1992.

Holmes, Burnham. *Basic Training: A Portrait of Today's Army.* New York: Four Winds Press, 1977.

Loring, Nancy H., ed. *Women in the United States Armed Forces: Progress and Barriers in the 1980s.* Chicago: Inter-University Seminar on Armed Forces and Society, 1984.

Mitchell, Brian. *Weak Link: The Feminization of the American Military.* Washington, D.C.: Regnery Publishing, 1989.

Mitchell, Brian. *Women in the Military: Flirting With Disaster.* Washington, D.C.: Regnery Publishing, 1998.

Morden, Col. Bettie J., USA (Ret). *The Women's Army Corps, 1945–1978.* Army Historical Series. Washington, D.C.: U.S. Army Center of Military History, 2001.

Pogue, Forrest C. *George C. Marshall: Organizer of Victory, 1943–1945.* New York: Viking Press, 1973.

Rogan, Helen. *Mixed Company: Women in the Modern Army.* New York: Putnam, 1981.

Rogers, Robin. "Combat Exclusion Promotes Widespread Discrimination in Society." In *Women in the Military.* Edited by Wekesser and Polesetsky. San Diego, Calif.: Greenhaven Press, 1991.

Rustad, Michael. *Women in Khaki: The American Enlisted Woman.* New York: Praeger Special Studies, 1982.

Schneider, Dorothy. *Sound Off: American Military Women Speak Out.* New York: E.P. Dutton, 1988.

Segal, Mady Wechsler. "The Argument for Female Combatants." In *Female Soldiers—Combatants or Noncombatants: Historical and Contemporary Perspectives.* Edited by Nancy Loring Goldman. Westport, Conn.: Greenwood Press, 1982.

Simon, Rita James, ed. *Women in the Military.* New Brunswick, N.J.: Transaction Publishers, Rutgers University, 2001.

Stiehm, Judith Hicks. *Arms and the Enlisted Woman.* Philadelphia: Temple University Press, 1989.

Treadwell, Mattie E. *U.S. Army in World War II: Special Studies—The Women's Army Corps.* Washington, D.C.: U.S. Army Center of Military History, 1954.

Tuten, Jeff M. "The Argument Against Female Combatants." In *Female Soldiers: Combatants or Noncombatants: Historical and Contemporary Perspectives.* Edited by Nancy Loring Goldman. Westport, Conn.: Greenwood Press, 1982.

Weigley, Russell F., ed. *History of the United States Army.* Bloomington: Indiana University Press, 1984.

Wekesser, Carol, and Matthew Polesetsky, eds. *Women in the Military: Current Controversies.* San Diego, Calif.: Greenhaven Press, 1991.

Articles and Papers

Ackmann, Martha. "Restricting Women's Military Role Hurts All," *Newsday* (8 April 2003).

Arney, Megan. "Military Report Admits: 'Sexual Harassment Exists Throughout Army.'" *The Militant* (28 September 1997).

Beck, Melinda, et al. "Our Women in the Desert." *Newsweek* (10 September 1990).

Cambanis, Thanassis. "U.S. Contests Lawsuit vs. Draft: Argues Against Including Women." *Boston Globe* (20 May 2003).

Crawley, Vince. "Women's Group Concerned About its Future." *Air Force Times* (24 December 2001).

Davey, Monica. "For 1,000 Troops, There Is No Going Home" *New York Times* (9 September 2004).

Devilbiss, M.C. "Women in the Army Policy Review: A Military Sociologist's Analysis." *Minerva: Quarterly Report on Women and the Military* (Fall 1983): 90–106.

Donnelly, Elaine. "Women in Combat: Time for a Review." *American Legion.* (July 2000).

Farrelly, Maura Jane. "Iraq War Renews Women in Combat Debate." *Iraq Crisis Bulletin.* (1 April 2003).

Finch, Mimi. "Women in Combat: One Commissioner Reports." *Minerva: Quarterly Report on Women and the Military* (Spring 1994).

Gabriel, Richard A. "Women in Combat: Two Views." *Army.* (March 1980).

Grossman, Robert J. "It's not Easy Being Green...and Female." *Human Resources.* (September 2001).

Hodges, Kathryn M. "Women in the Army Policy Review." *Minerva: Quarterly Report on Women and the Military* (22 June 1983): 86–89.

Infield, Tom. "Today's Army Differs from Gulf War Force." *Philadelphia Inquirer* (2 February 2003).

Ladine, Bret. "Army Unit to Bar Women." *Boston Globe.* (June 2002).

Ledford, Tranette. "Women at War." *Army Times.* (24 November 2003).

Maze, Rick. "Congress' Sole Woman Veteran Has Problems with DACOWITS' Plan." *Army Times.* (6 March 2002).

Moniz, Dave. "Army Revamps Training; Insurgent Tactics Prompt Changes." *USA Today* (6 January 2004).

Scarborough, Rowan. "Panel Queries Army's Plan for Women: Fears Change in Combat Rules." *Washington Times* (5 July 2001).

———. "Pentagon Urged to Separate Sexes in Basic Training: Groups Want Bush Promise Upheld." *Washington Times* (7 August 2001).

———. "Panel on Military Women in Peril: Rumsfeld Urged to Let It Expire." *Washington Times* (28 February 2002).

———. "U.S. Moves Women Away From Combat: Defense Panel to Shift From Combat Assignments to Readiness Issues." *Washington Times* (6 March 2002).

———. "Women Taken out of Army Squads: Policy Reversal Affects Recon Units." *Washington Times* (30 May 2002).

———. "Female Warriors Kept Off Ground for Special-Operations Missions." *Washington Times* (24 October 2001).

Small, Stephen. "Women in American Military History, 1776–1918." *Military Review* Vol. LXXVIII, No. 2 (March–April 1998): 101–104.

Spencer, Jack. "Why the Social Experiment of Gender-Integrated Training Has Failed." Heritage Foundation Executive Memorandum No. 758 (18 July 2001).

———. "Heritage Foundation Views 2001." (9 August 2001).

Winograd, Erin Q. "Army Opening Most Air Defense Artillery Slots to Female Soldiers." *Inside the Army.* (3 March 2003).

Yoest, Charmaine. "Booby Traps at the Pentagon." *Women's Quarterly* (Winter 2002).

Young, Anna M. "Army Women: Looking Toward an Uncertain Future (Again)." *Minerva: Quarterly Report on Women and the Military* (Spring 1984).

Internet and Broadcast Resources

Center for Military Readiness

Chastain, Jane. "So Much for a Kinder, Gentler Military," *World Net Daily* (6 May 2004).

Concerned Women for America. "Women Removed From Army RSTA Squadrons but Many Servicewomen Remain in Combat Positions," 5 June 2002.

Crumbo, Chuck. "Army May Revise Basic Training at Fort Jackson," www.thestate.com, South Carolina's Home Page, 7 January 2004.

Defense Technological Information Center

Independent Women's Forum

Kassandra Calhoun, Suite University Web site, 7 November 2000

Kirkwood, R. Cort. "What Kind of Nation Sends Women into Combat?" *Daily News-Record,* Harrisonburg, Va., 11 April 2003.

Metrakos, Rev. Aris P. "S&M Soldiers: How Sexual Politics Has Undermined the U.S. Military," *Orthodoxy Today*, 30 May 2004.

Morahan, Lawrence. "Women's Groups Blast 'Politically Correct' Pentagon Policies," Cybercast News Service, 31 January 2002.

Parker, Kathleen. "Jessica Lynch's Story Is About a Girl, Not a Soldier," www.townhall.com, 19 November 2003.

Pierce, Jason. "Coed Basic Training Hurts Military, Experts Say," Cybercast News Service, 25 July 2001.

Ryan, Joan. "Women and Uncle Sam," *San Francisco Chronicle,* 14 January 2003.

Thomas, Cal. "Sexual Politics and the Breakdown at Abu Ghraib," *Baltimore Sun*, 19 May 2004.

Thomson Gale: Free Resources

Vlahos, Kelly Beaucar. "CIA Recreating Military Might," *Fox News,* 19 October 2001.

Vlahos, Kelly Beaucar. "Pentagon Revamps Women's Military Panel," *Fox News*, 7 March 2002.

Yeagley, David. "Women Warriors," www.frontpagemagazine.com, 17 June 2002.

Miscellaneous Sources

Becraft, Carolyn. "Facts About Women in the Military." Women's Research and Education Institute, June 1990.

Blair, Anita. Interview with "About Face," Independent Women's Forum, 1 August 1999.

Center for Military Readiness. "New DACOWITS' Chair Named," 30 October 2002.

Donnelly, Elaine. Center for Military Readiness. *Army Gender-Integrated Basic Training: Summary of Relevant Findings and Recommendations: 1993–2002*, May 2003.

Fact Sheet, Women's Research and Education Institute, 1990.

Abbreviations and Acronyms

AHR	Annual Historical Review
AIT	advanced individual training
ARI	U.S. Army Research Institute for the Behavioral and Social Sciences
AVF	all-volunteer force
BCT	basic combat training
BIET	basic initial-entry training
CMH	Center of Military History
CSA	Chief of Staff, U.S. Army
DACOWITS	Defense Advisory Committee on Women in the Services
DCPC	Direct Combat Probability Coding system
DCSPER	Deputy Chief of Staff for Personnel
DoD	Department of Defense
DTIC	Defense Technological Information Center
ERA	Equal Rights Amendment
EWITA	Evaluation of Women in the Army study
FORSCOM	Army Forces Command
FY	fiscal year
GAO	General Accounting Office (federal government)
GIT	gender-integrated training
GOSC	General Officer Steering Committee
HQDA	Department of the Army headquarters
IET	initial entry training
IG	Inspector General
IPR	in-process review
MAX WAC	Women Content in Units Force Development Test
MEPSCAT	Military Enlistment Physical Strength Capacity Test
MILPERCEN	Military Personnel Center
MLRS	Multiple Launch Rocket System
MOS	military occupational specialty

NCO	noncommissioned officer
NORC	National Opinion Research Center at the University of Chicago
NOW	National Organization for Women
OPTEMPO	operational tempo
OSUT	One Station Unit Training program
OTEA	Army Operational Test and Evaluation Agency
PL	Public Law
POI	program of instruction
POW	prisoner of war
REFORGER	Return of Forces to Germany
REF WAC 77	Women Content in the Army–REFORGER 77 study
RMC	Royal Military College of Canada
RSTA	Reconnaissance, Surveillance, and Target Acquisition
SOF	special operations forces
TDA	temporary duty assignment
TOE	Tables of Organization and Equipment
TRADOC	Training and Doctrine Command
USA	U.S. Army
USAF	U.S. Air Force
USMA	U.S. Military Academy at West Point
USN	U.S. Navy
VMI	Virginia Military Institute
WAAC	Women's Army Auxiliary Corps
WAC	Women's Army Corps
WITA	Women in the Army study

Index

Mixed-Gender Basic Training

Walker, Mary Edwards, 2
War Department. *See* Department of the Army.
Warner, John, 85
Warrior ethos, 163, 164
Washington, General George, 2
Washington Post, 159
Washington Times, 143, 144, 150
Weak Link: The Feminization of the American Military (Mitchell), 94
Weinberger, Caspar, 64, 68
  and DACOWITS letter, 76–77
  on sexual harassment, 55
  on women in armed forces, 60
  See also Department of Defense.
West, Togo D., Jr., 115, 116, 124, 127, 129
  on mixed-gender training, 108, 120, 121
  on sexual harassment, 110–11, 112, 113, 114
  See also Senior Review Panel on Sexual Harassment.
Wetzel, Maj. Gen. Robert L., 62, 64
Weyand, General Frederick C., 43
White, Thomas, 145, 147, 148
White, Brig. Gen. Thomas E., 144, 147
Wickham, General John A., Jr., 77
Widnall, Sheila, 110
Wilson, Heather A., 152, 153
Wincup, G. Kim, 127
WITA Policy Review Group, 58, 62, 63, 64, 65, 66, 72. *See also* Direct Combat Probability Coding (DCPC) System; Military Enlistment Physical Strength Capacity Test (MEPSCAT).
Wolfowitz, Paul, 152, 153, 156
Womanpause, 58

Women in the Army (WITA), 38
Women in the Army Policy Review, 25
Women Content in the Army—REFORGER 77 (REF WAC 77), 41, 42
Women Content in Units Force Development Test (MAX WAC), 36–37, 41
Women in the Military Project, 144, 168
Women in Military Service for America Memorial, 131
Women's Armed Services Integration Act (PL 80-625), 4, 5n15, 11, 15, 131
Women's Army Auxiliary Corps (WAAC), 3
Women's Army Corps (WAC), 3, 4, 10, 69, 171
  abolition of, 34, 35, 42
  DCSPER on, 35
  fraternization issues in, 27–28
  marriage in, 20
  and pregnancy, 19, 20
  training, 42, 43, 47
Women's Equity Action League, 8
Women's Medical Specialist Corps, 4
Women's Memorial Foundation, 131
Women's Military Aviators organization, 86
Women's Program for the Army, 3
Women's Research and Education Institute, 88, 144
World Trade Center attack, 145
World War I, 2
World War II, 3–4

Yeomanettes, 2
Yerks, Lt. Gen. Robert G., 62

Zeltman, Brig. Gen. Ronald, 64